T0093774

# Concise Guide to Optimization Models and Methods

Xian Wen Ng

# Concise Guide to Optimization Models and Methods

## A Problem-Based Test Prep for Students

Springer

Xian Wen Ng
Singapore, Singapore

ISBN 978-3-030-84416-5 ISBN 978-3-030-84417-2 (eBook)
https://doi.org/10.1007/978-3-030-84417-2



This Springer imprint is published by the registered company Springer Nature Switzerland AG
The registered company address is: Gewerbestrasse 11, 6330 Cham, Switzerland

# Preface

Optimization involves finding the best solution to a system or process that is subject to constraints. It is highly valued in numerous fields and professions, from the scientist and engineer to the market analyst and regular business owner, since most real-life scenarios invariably require one to make the best decisions that would achieve desired outcomes, while keeping within any imposed limitations and constraints.

With an increasingly unpredictable global landscape with ever-evolving environmental conditions and other constraints, the ability to develop and apply systematic methods that efficiently optimize complex problems has become ever more relevant in order to respond and adapt well to changing circumstances.

In light of modern-day issues such as escalating energy costs and tighter resource availabilities, increasingly stringent environmental regulations, and stiffening competition in product quality and pricing, mastering optimization techniques has easily become one of the most important skillsets of the future. The optimization process quickly turns challenging with a large number of variables and dimensions to deal with, and so systematic methods to help one formulate models that would efficiently and accurately lead to optimized solutions hold the key to breaking down and solving some of the seemingly most onerous and convoluted problems.

This book is targeted at beginning students taking an introductory course in optimization, or entry-level practising engineers looking to hone their problem-solving skills and practise building optimization models. While this book focuses largely on problems with static constraints, it is worth noting that the subject of optimal control extends widely into optimizations of performance indices subject to a range of dynamic constraints, which are typical of real-world physical systems with dynamic control problems.

Written in problem-solution format, this book will be useful as supplementary reference to mainstream textbooks for students, as it serves as a guide in handling challenging problems commonly encountered in test and examinations. With comprehensive worked solutions and detailed explanations provided for each problem, students will be able to follow the thought process of problem-solving from start to finish, thereby hone their skills in applying abstract theoretical concepts to solving practical problems, a critical step to acing examinations. The mix of academic and real-world problems presented in this book will also help students tackling term projects or graduate-level optimization courses.

The balance of academic and practical examples in this book will help students develop skills in building optimization models and formulating solutions to linear, non-linear and convex programming problems. Students will become proficient, not only in tackling tests and examinations, but also in relating the significance of desktop problems to a larger real-world context.

Singapore, Singapore Xian Wen Ng

# Acknowledgements

My heartfelt gratitude goes to the team at Springer for their unrelenting support and professionalism throughout the publication process. Special thanks to Michael Luby and Brian Halm for your constant effort and attention towards making this publication possible. I am also deeply appreciative of the reviewers for my manuscript who had provided excellent feedback and numerous enlightening suggestions to help improve the book's contents.

Finally, I wish to thank my loved ones who have, as always, offered only patience and understanding throughout the process of making this book a reality.

# Contents

# About the Author

**X. W. Ng** graduated with First-Class Honours from the University of Cambridge, UK, with a master's degree in chemical engineering and Bachelor of Arts in 2011 and was subsequently conferred a Master of Arts in 2014. She was ranked second in her graduating class and was the recipient of a series of college scholarships, including the Samuel Taylor Marshall Memorial Scholarship, Thomas Ireland Scholarship and British Petroleum Prize in Chemical Engineering, for top performance in consecutive years of academic examinations. She was also one of the two students from Cambridge University selected for the Cambridge-Massachusetts Institute of Technology (MIT) exchange programme in chemical engineering, which she completed with honours. During her time at MIT, she was also a part-time tutor for junior classes in engineering and pursued other disciplines including economics, real estate development and finance at MIT and the John F. Kennedy School of Government, Harvard University. Upon graduation, she was elected by her college fellowship to the title of scholar, as a mark of her academic distinction.

Since graduation, she has been keenly involved in teaching across various academic levels. Her area of specialization includes mathematics, science and engineering topics. Some of her recent works include *Engineering Problems for Undergraduate Students* and *Pocket Guide to Rheology*, both of which were written in a similar problem-based format, specifically aimed at students taking engineering and related STEM courses at higher education and university levels. These books aim to sharpen students' problem-solving skills and put them in good stead for tests and examinations.

# Basic Concepts, Lagrangian Methods and Linear Programming Problems

**Abstract** This chapter builds a strong foundation in the understanding of the basic concepts and first principles behind how optimization works through problem formulation, and touches on the necessary conditions for minimization and maximization problems and what they mean. The concept of the Lagrangian method is introduced with detailed examples of its application. This chapter also includes examples of simple optimization problems involving only linear functions, which will provide beginner practice in problem formulation.

**Keywords** State variable · Supply and demand problems · Source and market problems · Transportation problem · Network problem · Network flow problem · Lagrangian function · Lagrangian multiplier · Profit maximization · Control variable · Convexity · Concavity · Global optimum · Global maxima · Global minima

## Problem 1

**Discuss briefly what optimal control problems are and their significance with reference to real-life scenarios. Explain why such problems may also be referred to as "dynamic optimization problems" or "infinite dimensional" optimization problems.**

## Solution 1

An optimal control problem (or dynamic optimization problem) is a problem that involves determining an optimized performance index (set by the objective function), with the goal to achieve a particular optimality criterion. Optimal control theory is widely adopted in various applications, such as in economics where it is used to model resource management and hence optimize its use.

In particular, infinite dimensional problems are studied in areas such as shape optimization and process optimization. For example, finding the optimum shape of



X. W. Ng, *Concise Guide to Optimization Models and Methods*,
https://doi.org/10.1007/978-3-030-84417-2_1

an engineering structure such that it can carry a desired amount of weight while using the least amount of materials to fabricate. Or, one could optimize a certain transportation process by finding the shortest path between two locations, subject to constraints due to terrain conditions.

Due to the constantly evolving nature of multiple state variables, optimal control problems are also referred to as 'infinite dimensional' or 'dynamic'. Infinite dimensional problems commonly produce an optimal solution that is not a discrete quantity, but instead a continuous quantity (for example a continuous function for the shape of an object), that cannot be determined by a finite number of degrees of freedom. Due to the relative complexity of infinite dimensional problems as compared to finite dimensional problems, methods from partial differential equations are usually required to solve them.

Dynamic optimization problems (or infinite dimensional problems) typically involve the following elements:

- An objective function that sets the performance index
- Control function trajectories
- Design parameters
- A free completion time (can be determined as a design parameter)
- Initial conditions (can be determined as a design parameter)
- General interior and final time constraints imposed at specific times within the integration horizon.

Note that in control problems, we often encounter terms such as "state variables" and "control variables". State variables describe the mathematical "state" of a dynamic system which is also the "state" that the decision-maker faces. Control variables are variables that the decision-maker can choose. Such variables are commonly constrained to being non-negative.

## Problem 2

**A construction project requires two types of concretes A and B, in volumes of 45,000 $m^3$ and 30,000 $m^3$ respectively. Concrete A costs *$10/$*$m^3$ while concrete B costs *$12/$*$m^3$. To produce concretes A and B, three different types of raw materials 1, 2 and 3 need to be extracted from their respective sources, then transported to a facility to be mixed together in specific compositions. The following data were given for the respective stages of the production of the two types of concrete.**

1. ***Extraction and transportation:***

**The amounts of raw materials available at their extraction sites are 25,000 $m^3$, 15,000 $m^3$ and 50,000 $m^3$ respectively. The cost (per $m^3$) for the extraction and transportation of raw materials 1, 2 and 3 are *$3*, *$3.50* and *$4* respectively.**

2. *Mixing and final formulation:*

**The limits on the composition of raw materials 1, 2 and 3 present in concretes A and B are shown below.**

| Concrete Type | Limits on raw material content (volume %) |
|---|---|
| A | 1 $\geq$ 35%; 2 + 3 $\leq$ 65% |
| B | 2 $\geq$ 25%; 3 $\leq$ 55%; 1 + 3 $\leq$ 75% |

**After mixing the three raw materials together, small amounts of additives are then added to produce the final products A and B. It is given that the cost of additives (per $m^3$) for concretes A and B are $2 and $3 respectively.**

(a) **Formulate a linear programming model for the concrete production. In your formulation, define all variables and constraints, as well as the objective function for profit maximization. Comment on any assumptions made.**
(b) **Briefly explain what a "standard transportation problem" in linear programming means. Show how the model in part a can be reformulated to have the same structure as a standard transportation problem. Also, transform the original objective function into the form of cost minimization, as required by a standard transportation problem.**

# Solution 2

**(a)**
In this problem, we shall assume that there is no mass (or volume) loss of raw materials 1, 2 and 3 throughout the concrete production process, e.g. during transfers or transitions between stages of production.

Model variable:
It is logical to first identify the variable in our linear programming model, given that we want to maximize profit from the sale of the concrete end-product. This variable should be the quantity that we can vary as we go about our optimization process to seek the optimal point that corresponds to profit maximization.

In this problem, we are looking for the specific compositions of raw materials 1, 2 and 3 in concretes A and B respectively, such that the profit gained from selling the concrete end-products will be the maximum. Therefore we can define our variable as $x_{ij}$ which is the amount or volume (in $m^3$) of raw material $i$ required to make concrete $j$. Subscript $i$ refers to the specific raw material and can be denoted as 1, 2 or 3, while subscript $j$ refers to the concrete product and can be denoted A or B.

$$\text{Variable } x_{ij} \begin{cases} i : 1, 2, 3 \\ j : A, B \end{cases}$$

Constraints on raw material availability:

Using the notation convention defined in our problem variable, let us now express the constraints on raw material availability as a set of linear equations or inequalities, as appropriate.

We are given the total amounts of each raw material available at their extraction sites, and we know that logically, we cannot extract more than what is available. Therefore the raw material availability constraints should be expressed as inequalities as shown below, whereby the stated amounts should not be exceeded.

$$x_{1A} + x_{1B} \leq 25{,}000 \cdots \boxed{1}$$

$$x_{2A} + x_{2B} \leq 15{,}000 \cdots \boxed{2}$$

$$x_{3A} + x_{3B} \leq 50{,}000 \cdots \boxed{3}$$

Constraints on amount of concrete required:

The construction project has a defined amount of concrete that it requires. This sets further constraints on our model, expressed as equalities as shown below:

For concrete A:

$$x_{1A} + x_{2A} + x_{3A} = 45{,}000 \cdots \boxed{4}$$

For concrete B:

$$x_{1B} + x_{2B} + x_{3B} = 30{,}000 \cdots \boxed{5}$$

Constraints on raw material compositions in concretes A and B:

Next, we note that for each type of concrete A and B, there are specific limits on the composition of raw material contained within. Hence we can express these constraints according to the specifications as listed in the table in the problem.

For concrete A:

$$x_{1A} \geq 0.35(x_{1A} + x_{2A} + x_{3A})$$

$$x_{2A} + x_{3A} \leq 0.65(x_{1A} + x_{2A} + x_{3A})$$

For concrete B:

$$x_{2B} \geq 0.25(x_{1B} + x_{2B} + x_{3B})$$

$$x_{3B} \leq 0.55(x_{1B} + x_{2B} + x_{3B})$$

$$x_{1B} + x_{3B} \leq 0.75(x_{1B} + x_{2B} + x_{3B})$$

Note that over here, we can do an intermediate simplification between constraints, by substituting constraint Eqs. ($\boxed{4}$) and ($\boxed{5}$) into the above expressions for raw material compositions.

This gives us the next 5 constraints ($\boxed{6}$), ($\boxed{8}$), ($\boxed{9}$) and ($\boxed{10}$) as follows:

For concrete A:

$$x_{1A} \geq 0.35(45,000)$$

$$x_{1A} \geq 15,750 \cdots \boxed{6}$$

$$x_{2A} + x_{3A} \leq 0.65(45,000)$$

$$x_{2A} + x_{3A} \leq 29,250 \cdots \boxed{7}$$

For concrete B:

$$x_{2B} \geq 0.25(30,000)$$

$$x_{2B} \geq 7,500 \cdots \boxed{8}$$

$$x_{3B} \leq 0.55(30,000)$$

$$x_{3B} \leq 16,500 \cdots \boxed{9}$$

$$x_{1B} + x_{3B} \leq 0.75(30,000)$$

$$x_{1B} + x_{3B} \leq 22,500 \cdots \boxed{10}$$

<u>Non-negativity constraints on physical quantities:</u>

In every problem modelling after a real-life scenario, it is good practice to always do a sense-check on the non-negativity constraint. This means that physical quantities such as volume, mass etc. would only be meaningful if they were non-negative numbers. We should also establish this constraint in our model to ensure our program is fully defined and hence able to produce meaningful solution(s).

$$x_{ij} \geq 0 \cdots \boxed{11}$$

Finally we write down our objective function $f(x_{ij})$ for profit maximization, subject to constraint equations and inequalities ($\boxed{1}$) to ($\boxed{11}$).

$$\max_{x_{ij}} f(x_{ij}) = \text{Concrete sales revenue} - \text{additives cost}$$

$$- \text{raw material extraction and transportation cost}$$

$$= 10(x_{1A} + x_{2A} + x_{3A}) + 12(x_{1B} + x_{2B} + x_{3B}) - 2(x_{1A} + x_{2A} + x_{3A}) - 3(x_{1B} + x_{2B} + x_{3B}) - 3(x_{1A} + x_{1B}) - 3.5(x_{2A} + x_{2B}) - 4(x_{3A} + x_{3B})$$

$$= 10(45,000) + 12(30,000) - 2(45,000) - 3(30,000) - 3(x_{1A} + x_{1B}) - 3.5(x_{2A} + x_{2B}) - 4(x_{3A} + x_{3B})$$

$$= 630,000 - 3(x_{1A} + x_{1B}) - 3.5(x_{2A} + x_{2B}) - 4(x_{3A} + x_{3B}), \quad \text{s.t.constraints } (1) \text{ to } (11)$$

**(b)**

A standard transportation problem in linear programming is an optimization problem with a linear objective function and linear constraints. It is often used to find the most efficient transportation routes that minimize total transportation cost. Hence, the model is commonly formulated as an objective function of transportation cost for minimization.

The general form of a linear program (LP) transportation problem is as shown:

$$\min \sum_{i=1}^{m} \sum_{j=1}^{n} c_{ij} x_{ij}$$

Subject to

$$\sum_{j=1}^{n} x_{ij} \le a_i, \quad for\ i = 1, 2 \ldots m$$

$$\sum_{i=1}^{m} x_{ij} \ge b_j, \quad for\ j = 1, 2 \ldots n$$

$$x_{ij} \ge 0, \quad for\ i = 1, 2 \ldots m\ and\ j = 1, 2 \ldots n$$

- $x_{ij}$ is the size of shipment from source $i$ to destination $j$. This is a set of $m \times n$ variables.
- The objective function is a linear total cost function, obtained by multiplying cost per unit of shipment $c_{ij}$ with the size of each shipment $x_{ij}$. This is summed over all $i$ and $j$ to account for all possible combinations of source-destination.

- There are constraints (functional constraint) on total supply at a particular source $i$ denoted $a_i$. The total outgoing shipment from source $i$ cannot exceed total supply at source $i$ or $a_i$.
- Similarly there are constraints (functional constraint) on total demand at destination $j$ denoted $b_j$. The total incoming shipment to destination $j$ cannot be less than the demand at destination $j$ or $b_j$.
- This gives an overall linear program with $m \times n$ decision variables, $m + n$ functional constraints, and $m \times n$ non-negative constraints.

Standard form of the transportation LP model:

When we have total supply equivalent to total demand, then $\sum_{i=1}^{m} a_i = \sum_{j=1}^{n} b_j$ and the model becomes simplified to its standard form, when inequality constraints become equalities.

$$\min \sum_{i=1}^{m} \sum_{j=1}^{n} c_{ij} x_{ij}$$

Subject to

$$\sum_{j=1}^{n} x_{ij} = a_i, \quad for\ i = 1, 2 \ldots m$$

$$\sum_{i=1}^{m} x_{ij} = b_j, \quad for\ j = 1, 2 \ldots n$$

$$x_{ij} \geq 0, \quad for\ i = 1, 2 \ldots m\ and\ j = 1, 2 \ldots n$$

In order to reformulate the problem in part a to have a structure similar to standard transportation problems, we will need to remove constraints ($\boxed{7}$) and ($\boxed{10}$).

$$x_{2A} + x_{3A} \leq 29{,}250 \cdots \boxed{7}$$

$$x_{1B} + x_{3B} \leq 22{,}500 \cdots \boxed{10}$$

Our original objective function was as follows:

$$\max_{x_{ij}} f(x_{ij}) = 630,000 - 3(x_{1A} + x_{1B}) - 3.5(x_{2A} + x_{2B}) - 4(x_{3A} + x_{3B})$$

We can ignore the constant term in optimization, hence the above is the same as maximizing the function as shown below.

$$\max_{x_{ij}} \left[-3(x_{1A}+x_{1B}) - 3.5(x_{2A}+x_{2B}) - 4(x_{3A}+x_{3B})\right]$$

Now, we can transform this original objective function for minimization by changing signs as shown below.

$$\begin{aligned} &\max_{x_{ij}} \left[-3(x_{1A}+x_{1B}) - 3.5(x_{2A}+x_{2B}) - 4(x_{3A}+x_{3B})\right] \\ =\ &\min_{x_{ij}} -\left[-3(x_{1A}+x_{1B}) - 3.5(x_{2A}+x_{2B}) - 4(x_{3A}+x_{3B})\right] \\ =\ &\min_{x_{ij}} \left[3(x_{1A}+x_{1B}) + 3.5(x_{2A}+x_{2B}) + 4(x_{3A}+x_{3B})\right] \end{aligned}$$

Our reformulated problem will therefore be a minimization of the function 3 $(x_{1A}+x_{1B})+3.5(x_{2A}+x_{2B})+4(x_{3A}+x_{3B})$ subject to the same set of constraints in part a except for constraints ($\boxed{7}$) and ($\boxed{10}$), as required by a standard transportation problem.

## Problem 3

**Due to the outbreak of a viral pandemic, a logistic company has been tasked to transport face masks in units of packets, through a distribution network as shown below comprising of 3 supply sources $\boxed{A}$, $\boxed{B}$ and $\boxed{C}$, as well as 4 demand markets $\boxed{A}$, $\boxed{B}$, $\boxed{C}$ and $\boxed{D}$. The arrows indicate transportation routes that link a source to a market, and the numbers above the arrows represent transportation cost in dollars ($) per packet. The numbers on the left and right columns indicate the available number of packets at each source and the required demand at each market respectively.**

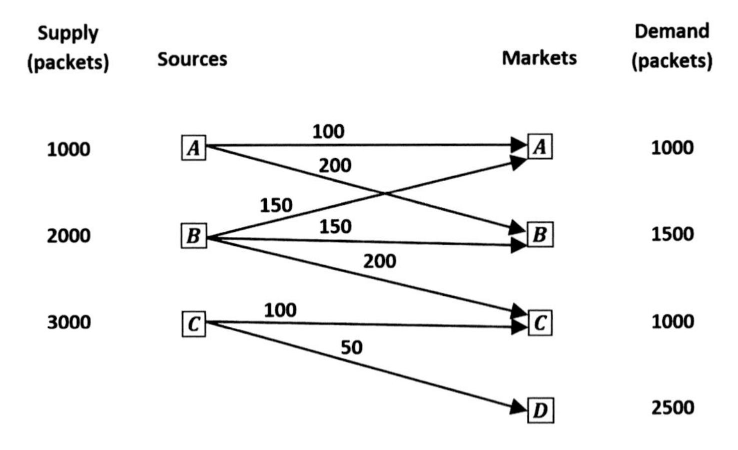

(a) **Formulate an optimisation model that minimizes transportation cost and identify the number of decision variables in your model.**

(b) **Show how some of the decision variables (identified in part a) can be easily calculated. Hence simplify the model in part a such that the new model has a reduced number of variables. How many constraints does this simplified model have?**

(c) **Show that the simplified model in part b can be further compacted into 2 equality constraints with 3 variables, and rewrite the objective function for this case. Calculate the optimum point, and determine the values for all the original variables and state the minimum transportation cost.**

(d) **How many variables would the model require if there was full connectivity from sources to markets? Assuming full connectivity, discuss what methods can be used to solve the optimization problem of minimizing transportation cost, and the possible constraints of some of these methods if any.**

## Solution 3

**(a)**

Since this problem involves minimization of transportation cost, the objective function for minimization would be the total transportation cost for the goods from all sources to all markets.

Note that in programming, it is easier for the model to process numbers instead of alphabets, hence we will convert the labels for sources $\boxed{A}$, $\boxed{B}$ and $\boxed{C}$ into $\boxed{1}$, $\boxed{2}$ and $\boxed{3}$ and do the same for the demand markets $\boxed{A}$ to $\boxed{D}$ which will be relabeled $\boxed{1}$ to $\boxed{4}$.

The objective function denoted $f$ below can now be expressed as follows, whereby the variable $x_{ij}$ represents the number of packets transported from source $i$ to market $j$:

$$\min f = 100x_{11} + 200x_{12} + 150x_{21} + 150x_{22} + 200x_{23} + 100x_{33} + 50x_{34}$$

This objective function is subject to constraints. We can first list down source balance and demand balance equations as shown.

<u>Source balance:</u>

$$1000 = x_{11} + x_{12} \cdots \boxed{i}$$

$$2000 = x_{21} + x_{22} + x_{23} \cdots \boxed{ii}$$

$$3000 = x_{33} + x_{34} \cdots \boxed{iii}$$

Demand balance:

$$1000 = x_{11} + x_{21} \cdots \boxed{iv}$$

$$1500 = x_{12} + x_{22} \cdots \boxed{v}$$

$$1000 = x_{23} + x_{33} \cdots \boxed{vi}$$

$$2500 = x_{34} \cdots \boxed{vii}$$

As with most physical quantities, we should establish non-negativity constraints to ensure the solution obtained are meaningful values.

Non-negativity:

$$x_{ij} \geq 0 \cdots \boxed{viii}$$

*where* $i = 1, 2, 3; j = 1, 2, 3, 4$

We can now conclude our optimization as a minimization of function *f* subject to constraints ($\boxed{i}$) to ($\boxed{viii}$).

From our model, we have a total of **7 decision variables** which are $x_{11}$, $x_{12}$, $x_{21}$, $x_{22}$, $x_{23}$, $x_{33}$ and $x_{34}$.

**(b)**

Out of the 7 decision variables identified in part a, we can easily calculate the values of 3 of them algebraically as shown below.

From Eq. ($\boxed{vii}$), we already have the value of one of the variables $x_{34}$.

$$x_{34} = 2500$$

Substituting the value of $x_{34}$ into the earlier Eq. ($\boxed{iii}$), we can find $x_{33}$.

$$x_{33} + x_{34} = 3000$$

$$x_{33} + 2500 = 3000$$

$$x_{33} = 500$$

Substituting the value of $x_{33}$ into the earlier Eq. ($\boxed{vi}$), we can also determine $x_{23}$.

$$x_{23} + x_{33} = 1000$$

$$x_{23} + 500 = 1000$$

$$x_{23} = 500$$

Hence we have easily solved for 3 variables $x_{34} = 2500$, $x_{33} = 500$ and $x_{23} = 500$.

We can derive a reduced (or simplified) model by substituting the known value of $x_{23}$ back into Eq. ($\boxed{ii}$) of our original model.

$$2000 = x_{21} + x_{22} + x_{23} \cdots \boxed{ii}$$

$$2000 = x_{21} + x_{22} + 500$$

$$1500 = x_{21} + x_{22} \cdots \boxed{ii*}$$

The earlier Eqs. ($\boxed{i}$), ($\boxed{iv}$) and ($\boxed{v}$) cannot be further simplified, so they are restated below.

$$1000 = x_{11} + x_{12} = \cdots \boxed{i}$$

$$1000 = x_{11} + x_{21} \cdots \boxed{iv}$$

$$1500 = x_{12} + x_{22} \cdots \boxed{v}$$

We now have a simplified model consisting of a reduced number (4 instead of 7) of variables $x_{11}$, $x_{12}$, $x_{21}$ and $x_{22}$, with four associated constraint Eqs. ($\boxed{ii*}$), ($\boxed{i}$), ($\boxed{iv}$) and ($\boxed{v}$).

**(c)**
We can substitute Eq. ($\boxed{ii*}$) into Eq. ($\boxed{iv}$)

$$1500 = x_{21} + x_{22} \cdots \boxed{iv}$$

$$x_{21} = 1500 - x_{22}$$

$$1000 = x_{11} + x_{21} = x_{11} + (1500 - x_{22})$$

$$500 = x_{22} - x_{11} \cdots \boxed{ix}$$

We can similarly substitute Eq. ($\boxed{i}$) into Eq. ($\boxed{v}$)

$$1000 = x_{11} + x_{12} \cdots \boxed{v}$$

$$x_{12} = 1000 - x_{11}$$

$$1500 = x_{12} + x_{22} = (1000 - x_{11}) + x_{22}$$

$$-500 = x_{11} - x_{22} \cdots \boxed{x}$$

At this point, we notice that Eqs. ($\boxed{ix}$) and ($\boxed{x}$) are the same, and therefore no new information is gained by retaining both of them, which means we should use just one of the two as shown below.

$$x_{11} = x_{22} - 500 \cdots \boxed{ix/x}$$

Substituting Eq. ($\boxed{ix/x}$) into the earlier Eq. ($\boxed{iv}$) we have,

$$x_{21} = 1500 - x_{22} = 1500 - x_{11} - 500$$
$$x_{21} = 1000 - x_{11}$$

What we have obtained above is the same as in Eq. ($\boxed{v}$), therefore we can rewrite the above as simply

$$x_{21} = x_{12} \cdots \boxed{xi}$$

Next, we can similarly substitute Eq. ($\boxed{ix/x}$) into the earlier Eq. ($\boxed{v}$) to obtain,

$$x_{12} = 1000 - x_{11} = 1000 - (x_{22} - 500)$$

$$x_{12} = 1500 - x_{22} \cdots \boxed{xii}$$

Our objective function can now be simplified with the new expressions of ($\boxed{ix/x}$), ($\boxed{xi}$) and ($\boxed{xii}$) obtained, as well as the earlier computed values for $x_{34} = 2500$, $x_{33} = 500$ and $x_{23} = 500$ as follows:

$$f = 100x_{11} + 200x_{12} + 150x_{21} + 150x_{22} + 200x_{23} + 100x_{33} + 50x_{34}$$
$$= 100x_{11} + 350x_{12} + 150x_{22} + 200x_{23} + 100x_{33} + 50x_{34}$$
$$= 100(x_{22} - 500) + 350(1500 - x_{22}) + 150x_{22} + 200(500) + 100(500) + 50(2500)$$

$$f = -100x_{22} + 750000$$

In minimizing the objective function $f$, we note that there is only one variable $x_{22}$ which has a negative coefficient. Therefore the maximum possible positive value of $x_{22}$ would give the minimum possible value of $f$.

$$\min f = \min(-100x_{22} + 750000) \rightarrow \max x_{22}$$

This means that the transportation route connecting supply source $\boxed{B}$ to demand market $\boxed{B}$ should be maximized in order to minimize transportation cost. The maximum value that the $\boxed{B}$ to $\boxed{B}$ route can transport can be obtained either by fully exhausting the supply limit of source $\boxed{B}$ or by fully fulfilling the demand limit of market $\boxed{B}$. The former is not possible since the supply limit of $\boxed{B}$ which is 2000 is greater than the demand limit of market $\boxed{B}$ which is 1500. Hence the latter scenario is the solution, which means the maximum value of $x_{22}$ is equivalent to the demand limit of market $\boxed{B}$ at 1500 packets. All of the demand for market $\boxed{B}$ will be completely fulfilled by source $\boxed{B}$ alone for transportation cost minimization.

Substituting the value of $x_{22} = 1500$ into the rest of the constraint equations, we can solve for all other variables in this problem.

$$x_{12} = x_{21} = 1500 - x_{22} = 1500 - 1500 = 0$$

$$x_{11} = x_{22} - 500 = 1500 - 500 = 1000$$

$$x_{23} = 500$$

$$x_{33} = 500$$

$$x_{34} = 2500$$

$$f_{optimum} = -100x_{22} + 750000 = -100(1500) + 750000 = 600000$$

The minimum transportation cost is hence found to be \$600,000.

**(d)**

The model would require 12 variables if there was full connectivity from 3 sources to 4 markets. This is because each source can supply to 4 markets, giving rise to 4 variables. For example for source 1 alone, there would be 4 variables $x_{11}$, $x_{12}$, $x_{13}$ and $x_{14}$. So with a total of 3 sources, there would be $3 \times 4 = 12$ variables.

The methods one can use to solve this problem under full connectivity will be the same as with reduced connectivity, and such methods include algebraic manipulation by hand and/or via the use of linear programming solvers.

For more complex problems involving a large number of variables (also referred to as larger dimensionality problems), then the problem would be difficult to solve algebraically by hand. This could arise if the problem consisted of many more sources and/or markets, or if the transportation network were fully connected instead of partially connected (more routes to consider, greater computational complexity). In such cases, a solver would be preferable.

## Problem 4

**An architect wishes to determine the maximum area he can obtain in his design of a rectangular wall, without changing the perimeter of the wall. He was told that he could determine an extremum for the area that the wall can have using Lagrange multipliers. Show how he can do so, and comment on whether this extremum in area is a maximum or minimum.**

## Solution 4

Since the objective here is to maximize the area of the rectangular wall subject to a fixed perimeter (denoted $P$), we can formulate the optimisation problem as follows, whereby $l$ and $w$ denote the length and width of the rectangle respectively. It is then straightforward to express area, $A$ of the wall as $A = lw$. Since we wish to maximize the value of $A$, $A$ is our objective function with variables $l$ and $w$.

$$\max_{l,w} A = \max_{l,w} (lw) \qquad s.t. 2l + 2w = P$$

We can form our Lagrangian function starting from the general form as shown below where $f$ is the objective function, $g_1 = 0$ represents equality constraints and $g_2 \geq 0$ represents inequality constraints if present.

$$\mathcal{L} = f + \lambda g_1 + \mu g_2$$

In this problem, $f = A = lw$, and $g_1 = 0 = 2l + 2w - P$. We do not have any inequality constraints hence the third term in the Lagrangian function above can be omitted. We arrive at the following Lagrangian function for this problem.

$$\mathcal{L} = lw + \lambda(2l + 2w - P)$$

In order to determine the constrained extremum, we need to set conditions for optimality which are:

$$\nabla_{l,w}\mathcal{L} = \begin{cases} \nabla_l \mathcal{L} = 0 \\ \nabla_w \mathcal{L} = 0 \end{cases} \quad \text{and} \quad \nabla_\lambda \mathcal{L} = 0$$

Differentiating the Lagrangian function with respect to $l$, $w$ and $\lambda$, and setting them to zero as above, we obtain:

$$\nabla_l \mathcal{L} = \frac{d\mathcal{L}}{dl} = w + 2\lambda = 0 \cdots \boxed{1}$$

$$\nabla_w \mathcal{L} = \frac{d\mathcal{L}}{dw} = l + 2\lambda = 0 \cdots \boxed{2}$$

$$\nabla_\lambda \mathcal{L} = 2l + 2w - P = 0 \cdots \boxed{3}$$

By rearranging Eq. ($\boxed{3}$), we obtain the following expression, which we can substitute back into Eq. ($\boxed{1}$).

$$w = \frac{P}{2} - l$$

$$w + 2\lambda = 0 \rightarrow \frac{P}{2} - l + 2\lambda = 0$$

From Eq. (2), we know that $l = -2\lambda$, which we can substitute into the equation above to obtain an expression for $\lambda$.

$$\frac{P}{2} - (-2\lambda) + 2\lambda = 0$$

$$\lambda = -\frac{P}{8}$$

$$l = w = -2\lambda = \frac{\mathrm{P}}{4}$$

The results above based on the Lagrangian conditions show that at an extremum of area, the length and width are equal which describes a square. Let us now confirm if this extremum is the constrained maximum, by checking if the objective function (area) is concave. If so, then area would be maximised at the stationary point.

We had the condition that perimeter is fixed, which we will substitute into the objective function $A$

$$2l + 2w = P \rightarrow l = \frac{P}{2} - w$$

$$A = lw = \left(\frac{P}{2} - w\right)w = w\left(\frac{P}{2}\right) - w^2$$

At the stationary point, $\frac{dA}{dw} = 0$ and this occurs at $w = \frac{P}{4}$ as found below. This result is consistent with our earlier results using the Lagrangian function.

$$\frac{dA}{dw} = \frac{P}{2} - 2w = 0$$

$$w = \frac{P}{4}$$

If the stationary point is a maximum for a concave function, then $\frac{d^2A}{dw^2} < 0$ which we confirm is the case as shown below.

$$\left.\frac{d^2A}{dw^2}\right|_{w=\frac{P}{4}} = -2 < 0$$

We can also plot the graph of function $A$ and show that it is fully concave with a unique global maximum. Note also that since area can only take non-negative values to be physically meaningful, the valid range of values for $w$ is $0 \leq w \leq \frac{P}{2}$.

$$A = 0 \rightarrow w = \frac{P}{2}, 0$$

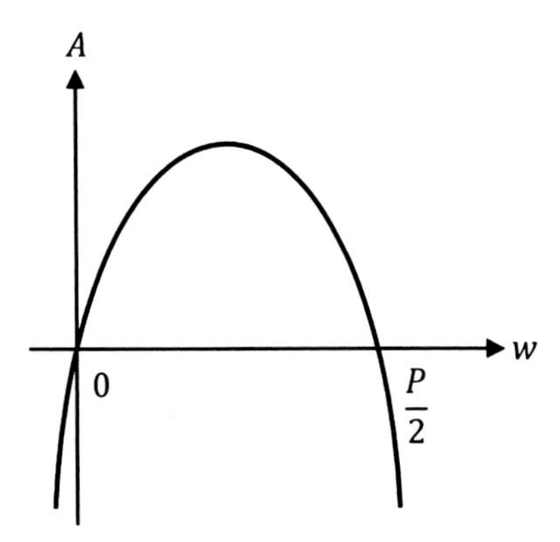


Therefore, we verify that the earlier-derived extremum point at $w = \frac{P}{4}$ is the constrained maximum. And the maximum area is indeed a square of equal sides, each measuring $\frac{P}{4}$.

## Problem 5

**Consider the following optimisation problem:**

$$\min_{x} \left(2x_1 + 3x_2 - 4x_3\right), \quad s.t. \begin{cases} x_1 + 3x_2 \leq 8 \\ -2x_1 + x_2 - 4x_3 = 1 \\ x_i \geq 0, i = 1, 2, 3 \end{cases}$$

**Show that the solution $x_1 = x_3 = 0, x_2 = 1$ is the optimal feasible point, based on the conditions required for optimality on the Lagrangian function.**

# Solution 5

The general Lagrangian function for a minimization problem is as follows, whereby $f(x)$ is the objective function and its minimization is subject to equality constraint $h(x) = 0$ and inequality constraints $g(x) \leq 0$. The parameters $\lambda$ and $\mu$ are Lagrange multipliers.

$$\mathcal{L}(x, \lambda, \mu) = f(x) + \lambda h(x) + \mu g(x)$$

Let us begin by expressing all the given constraints (equality and inequality constraints) in the required form and formulating our Lagrangian function as follows.

$$h(x) = -2x_1 + x_2 - 4x_3 - 1 = 0$$
$$g_1(x) = x_1 + 3x_2 - 8 \leq 0$$
$$g_2(x) = -x_1 \leq 0$$
$$g_3(x) = -x_2 \leq 0$$
$$g_4(x) = -x_3 \leq 0$$

$$\mathcal{L} = 3x_1 + 2x_2 - 7x_3 + \lambda(-2x_1 + x_2 - 4x_3 - 1) + \mu_1(x_1 + 3x_2 - 8) + \mu_2(-x_1) + \mu_3(-x_2) + \mu_4(-x_3)$$

Let us first use the given feasible solution $x_1 = x_3 = 0$, $x_2 = 1$ to eliminate any inactive inequality constraints from our Lagrangian function above. Since only $g_2$ and $g_4$ are active inequality constraints, $\mu_2, \mu_4 > 0$ and $\mu_1, \mu_3 = 0$.

$$g_1(x) = x_1 + 3x_2 - 8 = -5 \leq 0 \text{ (inactive)}$$
$$g_2(x) = -x_1 = 0 \text{ (active)}$$
$$g_3(x) = -x_2 \leq 0 \text{ (inactive)}$$
$$g_4(x) = -x_3 = 0 \text{ (active)}$$

Let us now update our Lagrangian function as shown,

$$\mathcal{L} = 3x_1 + 2x_2 - 7x_3 + \lambda(-2x_1 + x_2 - 4x_3 - 1) + \mu_2(-x_1) + \mu_4(-x_3)$$

To verify the optimality of the given solution, the feasible solution must satisfy the following necessary conditions for constrained optimality at $x^*$.

$$\left.\frac{d\mathcal{L}}{dx_i}\right|_{x^*} = 0, \quad \left.\frac{d\mathcal{L}}{d\lambda}\right|_{x^*} = 0, \quad \left.\frac{d\mathcal{L}}{d\mu_i}\right|_{x^*} = 0$$

Taking the above derivatives of $\mathcal{L}$, we have

$$\frac{d\mathcal{L}}{dx_1} = 3 - 2\lambda - \mu_2 = 0 \cdots \boxed{1}$$

$$\frac{d\mathcal{L}}{dx_2} = 2 + \lambda = 0 \cdots \boxed{2}$$

$$\frac{d\mathcal{L}}{dx_3} = -7 - 4\lambda - \mu_4 = 0 \cdots \boxed{3}$$

$$\frac{d\mathcal{L}}{d\lambda} = -2x_1 + x_2 - 4x_3 - 1 = 0 \cdots \boxed{4}$$

$$\frac{d\mathcal{L}}{d\mu_2} = -x_1 = 0 \cdots \boxed{5}$$

$$\frac{d\mathcal{L}}{d\mu_4} = -x_3 = 0 \cdots \boxed{6}$$

We have a total of 6 equations above, and 6 unknowns ($x_1$, $x_2$, $x_3$, $\lambda$, $\mu_2$, $\mu_4$). This means the system of equations is fully defined. From ($\boxed{5}$) and ($\boxed{6}$), we know $x_1 = x_3 = 0$. Substituting these values into Eq. ($\boxed{4}$), we obtain $x_2 = 1$.

$$0 + x_2 - 0 - 1 = 0$$

$$x_2 = 1$$

We can also compute the values of the Lagrange multipliers by solving the set of simultaneous Eqs. ($\boxed{1}$) to ($\boxed{3}$) as shown below.

$$2 + \lambda = 0 \rightarrow \lambda = -2$$

$$3 - 2\lambda - \mu_2 = 3 - 2(-2) - \mu_2 = 0 \rightarrow \mu_2 = 7 > 0$$

$$-7 - 4\lambda - \mu_4 = -7 - 4(-2) - \mu_4 = 0 \rightarrow \mu_4 = 1 > 0$$

We have shown above that the inequality constraints $g_2$ and $g_4$ are indeed active as their Lagrange multipliers are positive values greater than zero.

## Problem 6

**A furniture store is planning a production schedule for a new product over the next six weeks. The production capacities and costs, as well as expected demand for the product in each week are shown in the table below. For each unit of**

**product, other than production costs, there is also a storage cost for keeping the product in inventory from one week to the next. It is required that the inventory be cleared of all stock after six weeks (i.e. no stock in week seven). For week 1, it may be assumed that there is no stock in inventory.**

| | Week number | | | | | |
|---|---|---|---|---|---|---|
| | 1 | 2 | 3 | 4 | 5 | 6 |
| **Production capacity** | 120 | 120 | 180 | 200 | 180 | 170 |
| **Unit production cost, $** | 2.50 | 2.50 | 2.60 | 2.20 | 2.40 | 2.40 |
| **Unit storage cost, $** | – | 1.3 | 1.3 | 1.4 | 1.4 | 1.5 |
| **Demand** | 70 | 140 | 180 | 190 | 220 | 180 |

(a) **Formulate the optimisation problem, minimizing total cost for production.**
(b) **Given that the optimal production level in each week is given by the solution vector (110,120,180,200,180,170), find the amount of inventory stored each week, and compute the objective function value.**
(c) **Determine the value of the Lagrange multipliers corresponding to the constraints in the optimisation problem in part a evaluated at the solution given in part b.**
(d) **The store hopes to expand its production capacity of a single week, recommend with reasons the ideal week to do so. You may use your results in part c to justify your answer.**

## Solution 6

**(a)**
In this problem, our aim is to minimize the total cost of production. Therefore, our objective function $f$ will be a function describing total cost, and we would want to find the point at which $f$ is a minimum value (i.e. the constrained minimum).

$$\min f$$

In order to express $f$ in a useful expression, we need to identify our optimisation variables. They comprise of the production level (number of units of product made) in each week denoted $x_i$, and the inventory level (number of units of product in storage) in each week denoted $I_i$, where $i = 1, 2, \ldots 6$ in both cases for the 6-week period.

We can now express our cost function $f$ as follows, using the data given in the problem. Note that there is no inventory in the first week, hence $I_1 = 0$.

$$\min f = \min(2.5x_1 + 2.5x_2 + 2.6x_3 + 2.2x_4 + 2.4x_5 + 2.4x_6 + 1.3I_2 + 1.3I_3) \times (+1.4I_4 + 1.4I_5 + 1.5I_6)$$

Next, we will establish all the constraints that this optimisation is subject to. First, we note that there are limitations on production capacity as given in the problem. We can express them as a set of inequality constraints as shown below.

$$x_1 \leq 120 \cdots \boxed{1}$$

$$x_2 \leq 120 \cdots \boxed{2}$$

$$x_3 \leq 180 \cdots \boxed{3}$$

$$x_4 \leq 200 \cdots \boxed{4}$$

$$x_5 \leq 180 \cdots \boxed{5}$$

$$x_6 \leq 170 \cdots \boxed{6}$$

Next, we note that in order to have physically meaningful solutions, we should set the non-negativity constraint on inventory and production levels.

$$x_1, x_2, x_3, x_4, x_5, x_6 \geq 0 \cdots \boxed{7}$$

$$I_1, I_2, I_3, I_4, I_5, I_6, I_7 \geq 0 \cdots \boxed{8}$$

Now let us consider any relevant material balances. As we have inventory that is changing from week to week, we can correlate inventory level in a particular week $i$, $I_i$ to that in the immediate next week or the $(i + 1)^{th}$ week, $I_{i+1}$. The inventory in week $i$ (amount found in storage) added on to the amount of new product made in the same week $i$, and further subtracted by the amount of product removed by demand, would give the final inventory that will flow over to the next $(i + 1)^{th}$ week, and this value will then be $I_{i+1}$. If we denote the demand in week $i$ as $D_i$, we arrive at the following expression.

$$I_{i+1} = I_i + x_i - D_i$$

We should also note that there is no inventory in week 1, as well as after the sixth week, i.e. no inventory in week 7. Therefore, $I_1 = I_7 = 0$. We can now list down the inventory expressions for the weeks as follows.

$$I_1 = 0 \cdots \boxed{9}$$

$$I_2 = I_1 + x_1 - D_1 = 0 + x_1 - 70$$

$$I_2 = x_1 - 70 \cdots \boxed{10}$$

$$I_3 = I_2 + x_2 - D_2$$

$$I_3 = I_2 + x_2 - 140 \cdots \boxed{11}$$

$$I_4 = I_3 + x_3 - D_3$$

$$I_4 = I_3 + x_3 - 170 \cdots \boxed{12}$$

$$I_5 = I_4 + x_4 - D_4$$

$$I_5 = I_4 + x_4 - 180 \cdots \boxed{13}$$

$$I_6 = I_5 + x_5 - D_5$$

$$I_6 = I_5 + x_5 - 220 \cdots \boxed{14}$$

$$I_7 = I_6 + x_6 - D_6 = 0$$

$$I_6 + x_6 - 180 = 0 \cdots \boxed{15}$$

Finally, our optimisation problem formulation can be written as the minimization of our objective function subject to all constraints, i.e. ($\boxed{1}$) to ($\boxed{15}$) as shown below.

$$\min (2.5x_1 + 2.5x_2 + 2.6x_3 + 2.2x_4 + 2.4x_5 + 2.4x_6 + 1.3I_2 + 1.3I_3) \\ \times (+1.4I_4 + 1.4I_5 + 1.5I_6) \text{ s.t. } \boxed{1} \text{ to } \boxed{15}$$

**(b)**

Since we are given the solution vector comprising values of $x_i$ for each week that would achieve minimum cost, we can substitute these values into our earlier expressions for inventory as shown below:

$$x^* = (x_1, x_2, x_3, x_4, x_5, x_6) = (110, 120, 180, 200, 180, 170)$$

$$I_1 = 0$$

$$I_2 = x_1 - 70 = 110 - 70$$

$$I_2 = 40$$

$$I_3 = I_2 + x_2 - 140 = 40 + 120 - 140$$

$$I_3 = 20$$

$$I_4 = I_3 + x_3 - 170 = 20 + 180 - 170$$
$$I_4 = 30$$
$$I_5 = I_4 + x_4 - 180 = 30 + 200 - 180$$
$$I_5 = 50$$
$$I_6 = I_5 + x_5 - 220 = 50 + 180 - 220$$
$$I_6 = 10$$
$$I_7 = 10 + x_6 - 180 = 10 + 170 - 180 = 0$$
$$I_7 = 0$$

The value of the objective function at this constrained minimum can then be calculated by substituting the values of inventory and production level as shown below. We find that the minimum total cost is $2740.

$$f(x) = 2.5x_1 + 2.5x_2 + 2.6x_3 + 2.2x_4 + 2.4x_5 + 2.4x_6 + 1.3I_2 + 1.3I_3 + 1.4I_4 + 1.4I_5 + 1.5I_6$$

$$\begin{aligned} f(x^*) &= 2.5(110) + 2.5(120) + 2.6(180) + 2.2(200) + 2.4(180) + 2.4(170) \\ &\quad + 1.3(50) + 1.3(40) + 1.4(60) + 1.4(90) + 1.5(60) \\ &= 2740 \end{aligned}$$

**(c)**

We can first form our Lagrangian function $\mathcal{L}(x, I, \lambda, \mu)$ using the objective function and all established constraints (equalities and inequalities). The general form of the Lagrangian function for a minimization problem is as follows.

$$\mathcal{L}(x, I, \lambda, \mu) = f(x, I) + \lambda h(x, I) + \mu g(x, I) \quad s.t. h(x, I) = 0 \quad g(x, I) \le 0$$

We note our earlier constraints ($\boxed{1}$) to ($\boxed{6}$) are inequality constraints in the required form $g(x, I) \le 0$ as follows:

$$x_1 - 120 \le 0 \cdots \boxed{1}$$
$$x_2 - 120 \le 0 \cdots \boxed{2}$$
$$x_3 - 180 \le 0 \cdots \boxed{3}$$
$$x_4 - 200 \le 0 \cdots \boxed{4}$$
$$x_5 - 180 \le 0 \cdots \boxed{5}$$
$$x_6 - 170 \le 0 \cdots \boxed{6}$$

At the constrained minimum point $x^*$ given by $(x_1, x_2, x_3, x_4, x_5, x_6) =$ (110,120,180,200,180,170), we note that inequalities ($\boxed{2}$) to ($\boxed{6}$) are active as they are fulfilled as equalities, therefore their associated Lagrange multipliers will be $\mu_2$, $\mu_3$, $\mu_4$, $\mu_5$, $\mu_6 > 0$ and the associated constraint functions will appear in the Lagrangian function. On the other hand, constraint ($\boxed{1}$) is inactive as it is fulfilled as an inequality at the constrained point, hence its Lagrange multiplier $\mu_1 = 0$ and this constraint function will "drop out" of the Lagrangian function.

As for constraints ($\boxed{7}$) and ($\boxed{8}$), we can again rewrite them in the required form for inequalities as follows.

$$-x_1, -x_2, -x_3, -x_4, -x_5, -x_6 \leq 0 \cdots \boxed{7}$$

$$-I_1, -I_2, -I_3, -I_4, -I_5, -I_6, -I_7 \leq 0 \cdots \boxed{8}$$

We check again that at the constrained minimum, the inequality constraint ($\boxed{7}$) is inactive since all values of $x^* > 0$ or in other words $-x^* < 0$ (fulfilled as an inequality), therefore the associated Lagrange multipliers for ($\boxed{7}$) will all be zero.

As for constraint ($\boxed{8}$), our results in part b showed that at the solution, the values of $I^*$ are $(I_1, I_2, I_3, I_4, I_5, I_6, I_7) = (0{,}40{,}20{,}30{,}50{,}10{,}0)$ which means that only $-I_1, -I_7 \leq 0$ are active inequality constraints as they are fulfilled as equalities ($-I_1 = -I_7 = 0$). We can denote their associated Lagrange multipliers as $\mu_{8,\ 1}$ for $I_1$ and $\mu_{8,\ 7}$ for $I_7$ respectively whereby $\mu_{8,\ 1}, \mu_{8,\ 7} > 0$.

The rest of the inequalities, $-I_2, -I_3, -I_4, -I_5, -I_6 \leq 0$ are inactive since they fulfil $-I_2, -I_3, -I_4, -I_5, -I_6 < 0$, therefore their Lagrange multipliers are zero and these inequality functions will "drop out" of the Lagrangian function.

Now let us consider our final set of constraints, the equality constraints from ($\boxed{9}$) to ($\boxed{15}$). We can re-express them in the required form $h(x, I) = 0$ as shown below. Since equality constraints always have to be active at the constrained optimum, they are all required to be included in the Lagrangian function with an associated non-zero Lagrangian multiplier denoted $\lambda_9, \lambda_{10}, \lambda_{11}, \lambda_{12}, \lambda_{13}, \lambda_{14}, \lambda_{15} \neq 0$.

$$I_1 = 0 \cdots \boxed{9}$$

$$x_1 - 70 - I_2 = 0 \cdots \boxed{10}$$

$$I_2 + x_2 - 140 - I_3 = 0 \cdots \boxed{11}$$

$$I_3 + x_3 - 170 - I_4 = 0 \cdots \boxed{12}$$

$$I_4 + x_4 - 180 - I_5 = 0 \cdots \boxed{13}$$

$$I_5 + x_5 - 220 - I_6 = 0 \cdots \boxed{14}$$

$$I_6 + x_6 - 180 = 0 \cdots \boxed{15}$$

We are now ready to form our Lagrangian function of the general form below.

$$\mathcal{L}(x, I, \lambda, \mu) = f(x, I) + \lambda h(x, I) + \mu g(x, I)$$

By considering only active constraints defined as those that have non-zero multipliers, we have the following Lagrangian function for our problem.

$$\mathcal{L} = 2.5x_1 + 2.5x_2 + 2.6x_3 + 2.2x_4 + 2.4x_5 + 2.4x_6 + 1.3I_2 + 1.3I_3 + 1.4I_4 + 1.4I_5 + 1.5I_6$$

$$+\mu_2(x_2 - 120) + \mu_3(x_3 - 180) + \mu_4(x_4 - 200) + \mu_5(x_5 - 180) + \mu_6(x_6 - 170) + \mu_{8,1}(-I_1)$$

$$+\mu_{8,7}(-I_7) + \lambda_9(I_1) + \lambda_{10}(x_1 - 70 - I_2) + \lambda_{11}(I_2 + x_2 - 140 - I_3) + \lambda_{12}(I_3 + x_3 - 170 - I_4)$$

$$+\lambda_{13}(I_4 + x_4 - 180 - I_5) + \lambda_{14}(I_5 + x_5 - 220 - I_6) + \lambda_{15}(I_6 + x_6 - 180)$$

We can simplify this equation by noting that $I_1 = I_7 = 0$ at the constrained optimum, which allows us to remove the 3 terms "$\mu_{8,\ 1}(-I_1)$", "$\mu_{8,\ 7}(-I_7)$" and "$\lambda_9(I_1)$" from the above equation.

$$\mathcal{L} = 2.5x_1 + 2.5x_2 + 2.6x_3 + 2.2x_4 + 2.4x_5 + 2.4x_6 + 1.3I_2 + 1.3I_3 + 1.4I_4 + 1.4I_5 + 1.5I_6$$

$$+\mu_2(x_2 - 120) + \mu_3(x_3 - 180) + \mu_4(x_4 - 200) + \mu_5(x_5 - 180) + \mu_6(x_6 - 170) + \lambda_{10}(x_1 - 70 - I_2)$$

$$+\lambda_{11}(I_2 + x_2 - 140 - I_3) + \lambda_{12}(I_3 + x_3 - 170 - I_4) + \lambda_{13}(I_4 + x_4 - 180 - I_5)$$

$$+\lambda_{14}(I_5 + x_5 - 220 - I_6) + \lambda_{15}(I_6 + x_6 - 180)$$

We can now set our conditions for optimality, requiring that the derivative with respect to the variables, $\nabla \mathcal{L}_{x,I} = 0$ and the derivative with respect to the multipliers, $\nabla \mathcal{L}_{\lambda,\mu} = 0$.

For $\nabla \mathcal{L}_{x,I} = 0$, we have the following:

$$\frac{d\mathcal{L}}{dx_1} = 2.5 + \lambda_{10} = 0 \rightarrow \lambda_{10} = -2.5 \cdots \boxed{*}$$

$$\frac{d\mathcal{L}}{dx_2} = 2.5 + \mu_2 + \lambda_{11} = 0$$

$$\frac{d\mathcal{L}}{dx_3} = 2.6 + \mu_3 + \lambda_{12} = 0$$

$$\frac{d\mathcal{L}}{dx_4} = 2.2 + \mu_4 + \lambda_{13} = 0$$

$$\frac{d\mathcal{L}}{dx_5} = 2.4 + \mu_5 + \lambda_{14} = 0$$

$$\frac{d\mathcal{L}}{dx_6} = 2.4 + \mu_6 + \lambda_{15} = 0$$

$$\frac{d\mathcal{L}}{dI_2} = 1.3 - \lambda_{10} + \lambda_{11} = 0$$

$$\frac{d\mathcal{L}}{dI_3} = 1.3 - \lambda_{11} + \lambda_{12} = 0$$

$$\frac{d\mathcal{L}}{dI_4} = 1.4 - \lambda_{12} + \lambda_{13} = 0$$

$$\frac{d\mathcal{L}}{dI_5} = 1.4 - \lambda_{13} + \lambda_{14} = 0$$

$$\frac{d\mathcal{L}}{dI_6} = 1.5 - \lambda_{14} + \lambda_{15} = 0$$

We can substitute the value of $\lambda_{10}$ we found in ($\boxed{*}$) into the equation for $\frac{d\mathcal{L}}{dI_2} = 0$ in order to obtain the value of $\lambda_{11}$.

$$\frac{d\mathcal{L}}{dI_2} = 1.3 - \lambda_{10} + \lambda_{11} = 1.3 + 2.5 + \lambda_{11} = 0$$

$$\lambda_{11} = -3.8$$

We can continue this process of substituting the derived value of $\lambda_{11}$ into the next equation $\frac{d\mathcal{L}}{dI_3} = 0$ to find $\lambda_{12}$, and so on.

$$\frac{d\mathcal{L}}{dI_3} = 1.3 - \lambda_{11} + \lambda_{12} = 1.3 + 3.8 + \lambda_{12} = 0$$

$$\lambda_{12} = -5.1$$

$$\frac{d\mathcal{L}}{dI_4} = 1.4 - \lambda_{12} + \lambda_{13} = 1.4 + 5.1 + \lambda_{13} = 0$$

$$\lambda_{13} = -6.5$$

$$\frac{d\mathcal{L}}{dI_5} = 1.4 - \lambda_{13} + \lambda_{14} = 1.4 + 6.5 + \lambda_{14} = 0$$

$$\lambda_{14} = -7.9$$

$$\frac{d\mathcal{L}}{dI_6} = 1.5 - \lambda_{14} + \lambda_{15} = 1.5 + 7.9 + \lambda_{15} = 0$$

$$\lambda_{15} = -9.4$$

Substituting these derived values for $\lambda$ back into our earlier equations $\frac{d\mathcal{L}}{dx_2}$ to $\frac{d\mathcal{L}}{dx_6}$, we can find the values of $\mu_2$ to $\mu_6$ as well.

$$\frac{d\mathcal{L}}{dx_2} = 2.5 + \mu_2 + \lambda_{11} = 2.5 + \mu_2 - 3.8 = 0$$

$$\mu_2 = 1.3 > 0$$

$$\frac{d\mathcal{L}}{dx_3} = 2.6 + \mu_3 + \lambda_{12} = 2.6 + \mu_3 - 5.1 = 0$$

$$\mu_3 = 2.5 > 0$$

$$\frac{d\mathcal{L}}{dx_4} = 2.2 + \mu_4 + \lambda_{13} = 2.2 + \mu_4 - 6.5 = 0$$

$$\mu_4 = 4.3 > 0$$

$$\frac{d\mathcal{L}}{dx_5} = 2.4 + \mu_5 + \lambda_{14} = 2.4 + \mu_5 - 7.9 = 0$$

$$\mu_5 = 5.5 > 0$$

$$\frac{d\mathcal{L}}{dx_6} = 2.4 + \mu_6 + \lambda_{15} = 2.4 + \mu_6 - 9.4 = 0$$

$$\mu_6 = 7 > 0$$

We have hence derived all the values of our Lagrange multipliers. We can reaffirm that the optimal constrained solution to this problem exists at the condition $\nabla \mathcal{L}_{\lambda,\mu} = 0$ as follows:

$$\frac{d\mathcal{L}}{d\mu_2} = x_2 - 120 = 0 \rightarrow x_2 = 120$$

$$\frac{d\mathcal{L}}{d\mu_3} = x_3 - 180 = 0 \rightarrow x_3 = 180$$

$$\frac{d\mathcal{L}}{d\mu_4} = x_4 - 200 = 0 \rightarrow x_4 = 200$$

$$\frac{d\mathcal{L}}{d\mu_5} = x_5 - 180 = 0 \rightarrow x_5 = 180$$

$$\frac{d\mathcal{L}}{d\mu_6} = x_6 - 170 = 0 \rightarrow x_6 = 170$$

$$\frac{d\mathcal{L}}{d\lambda_{10}} = x_1 - 70 - I_2 = 0 \rightarrow 110 - 70 - 40 = 0$$

$$\frac{d\mathcal{L}}{d\lambda_{11}} = I_2 + x_2 - 140 - I_3 = 0 \rightarrow 40 + 120 - 140 - 20 = 0$$

$$\frac{d\mathcal{L}}{d\lambda_{12}} = I_3 + x_3 - 170 - I_4 = 0 \rightarrow 20 + 180 - 170 - 30 = 0$$

$$\frac{d\mathcal{L}}{d\lambda_{13}} = I_4 + x_4 - 180 - I_5 = 0 \rightarrow 30 + 200 - 180 - 50 = 0$$

$$\frac{d\mathcal{L}}{d\lambda_{14}} = I_5 + x_5 - 220 - I_6 = 0 \rightarrow 50 + 180 - 220 - 10 = 0$$

$$\frac{d\mathcal{L}}{d\lambda_{15}} = I_6 + x_6 - 180 = 0 \rightarrow 10 + 170 - 180 = 0$$

The above set of results confirm that $\nabla \mathcal{L}_{\lambda,\mu} = 0$ holds true at the constrained optimum, which means all active constraints will be satisfied as required.

**(d)**

Lagrange multipliers of constraint functions measure the "sensitivity" of the objective function to the constraint. In general, for a Lagrangian function of the form $\mathcal{L}(x, \lambda) = f(x) + \lambda h(x)$ whereby the constraint $h(x) = 0$ is active at the constrained optimum, the Lagrange multiplier $\lambda$ can be expressed as follows, which shows this "sensitivity" effect mathematically.

$$\lambda = -\frac{\partial f}{\partial h}\bigg|_{x^*}$$

Therefore, by observing the relative values of multipliers for different constraints, we are able to deduce which constraint(s) have greater effect on affecting the objective function. In other words, we can identify the constraint that would have the most impact on total cost if it has the largest associated multiplier. Any changes to this constraint will therefore also affect total cost most.

Looking at our set of constraints for this problem, we first note that only the capacity constraints on production levels are flexible to change via redesign of the store's operations. Other constraints such as non-negativity constraints and material balance constraints are less amenable to changes, as they describe more immutable aspects of the system (e.g. physical laws).

Let us examine the set of inequality constraints related to production capacity. The sensitivity of the inequality constraints in terms of multiplier $\mu$ can be expressed as shown below.

$$\mu = -\frac{\partial f}{\partial g}\bigg|_{x^*}, \quad g \leq 0$$

Note that the function $g$ is the production capacity constraint for a particular week, denoted $C_i$ below for an arbitrary week $i$.

$$x_i \leq C_i$$

$$g_i = x_i - C_i$$
$$g_i \leq 0, x_i - C_i \leq 0$$

From our answer in part c, we note that the Lagrange multipliers for the capacity constraints at the constrained minimum are as follows.

$$\mu_2 = 1.3$$
$$\mu_3 = 2.5$$
$$\mu_4 = 4.3$$
$$\mu_5 = 5.5$$
$$\mu_6 = 7$$

The multiplier that has the largest value is $\mu_6$. This implies that if we expanded the production capacity of week 6 i.e. $C_6$, it would result in the largest reduction in the value of the objective function at the constrained minimum.

The production capacity for week 6 can also be said to have the most impact in streamlining the store's operations since it reduces total cost most, as compared to the production capacity constraints in other weeks. Week 6 is therefore the most ideal week to expand production capacity.

$$\mu_6 = -\left.\frac{\partial f}{\partial g_6}\right|_{x^*}$$

If we expand production capacity in week 6, $C_6$ decreases (expansion in capacity due to reduction in the limit on capacity), then $g_6$ increases since $g_6 = x_6 - C_6$. This increase in $g_6$, if small, can be expressed as $\Delta g_6$ whereby

$$\Delta g_6 > 0, \Delta g_6 = g_{final} - g_{initial}$$

Due to the large value of the multiplier $\mu_6$, this would cause a large reduction in the value of $f$, for non-negative values of $\mu_6$,

$$\mu_6 \cong -\frac{\Delta f}{\Delta g_6}$$
$$\Delta g_6 > 0 \rightarrow -\Delta f > 0$$
$$-\Delta f = -\left(f_{final} - f_{initial}\right) = f_{initial} - f_{final} > 0 \text{ (i.e.decrease in} f)$$

# Non-linear Programming Problems with Constraints and Euler's Methods

**Abstract** This chapter introduces non-linear problems which require the consideration of various constraints in problem formulation. The worked solutions provided will guide students through problem solving techniques and train their ability to interpret complicated systems and systematically cast them into a concise set of objectives and parameters. The problems in this chapter include more rigorous methodologies such as discretization schemes, and Euler's methods that would set the stage for more complex problems that require computer solvers.

**Keywords** Reaction process optimization · Arrhenius equation · Reaction kinetics · Non-negativity constraint · Discretization scheme · Discrete elements · Blending problem · Mixing problem · Bilinear terms · Euler's method · Network flow · Forward difference · Backward difference · Grid based method · Implicit method

## Problem 7

**Two different blended vegetable oil products, X and Y are produced by mixing together three different types of oils: palm oil, soybean oil and sunflower oil respectively. The mixing and blending flow sheet is shown below.**

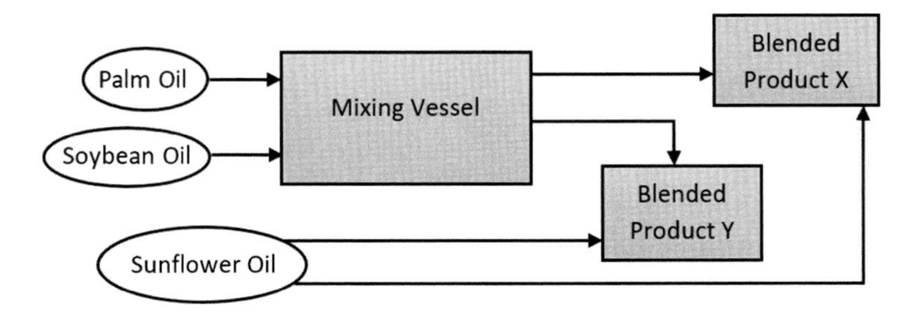


**The percentages of saturated fatty acids in each of the 3 oils (palm, soybean and sunflower) as well as the 2 blended oil products X and Y are shown in the**



X. W. Ng, *Concise Guide to Optimization Models and Methods*,
https://doi.org/10.1007/978-3-030-84417-2_2

**table below. The selling prices of the blended oil products are also included in the table below.**

| | Percentage of saturated fatty acids (%) | Selling Price ($ per volume unit) |
|---|---|---|
| Palm oil | 2.5 | 50 |
| Soybean oil | 0.9 | 150 |
| Sunflower oil | 1 | 100 |
| Blended oil, X | No more than 2 | 80 |
| Blended oil, Y | No more than 1.2 | 140 |

(a) **Given that the purchase order limit for the blended oil product X is 120 volume units while that for product Y is 250 volume units, formulate the problem as a non-linear programming model for profit maximisation. You may assume that all the oils have the same density.**
(b) **Comment on whether this problem is a convex or non-convex optimisation problem.**
(c) **Comment on whether this problem can be changed to a linear programming problem and show how it can be done if so.**
(d) **An engineer devised a solution as shown below whereby stream volumes (in volume units) are indicated above the respective stream arrows. Comment on whether this solution is viable, and determine value of the objective function.**

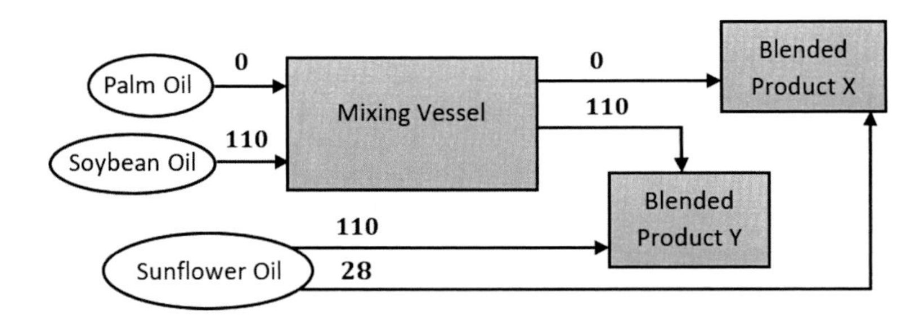


## Solution 7

**(a)**
In order to formulate a non-linear programming model for profit maximization, we need to first figure out our constraints, as well as our objective function. The objective function should spell out mathematically, what we aim to achieve with the optimization model, in this case, the objective is to maximize profit.

Let us first identify our constraints. One important constraint to consider, especially in material flow problems, is mass balance, since mass conservation is universal and is therefore one of the most commonly required constraints in formulating such models. Similarly we need to think about energy balance when we encounter energy (or enthalpy/heat) flow problems.

We can begin by drawing a control volume in dotted lines around the mixing vessel and performing a mass balance for this control volume on the oil content. We have denoted the mixing vessel as "M", and the volume units of oil are annotated as follows:

- $x_{p,\ M}$ refers to the number of volume units of palm oil entering the mixing vessel (M).
- $x_{sb,\ M}$ refers to the number of volume units of soybean oil entering the mixing vessel (M).
- $x_{M,\ X}$ refers to the number of volume units of oil in the outlet of mixing vessel (M) and entering the blending vessel for product X.
- $x_{M,\ Y}$ refers to the number of volume units of oil in the outlet of mixing vessel (M) and entering the blending vessel for product Y.

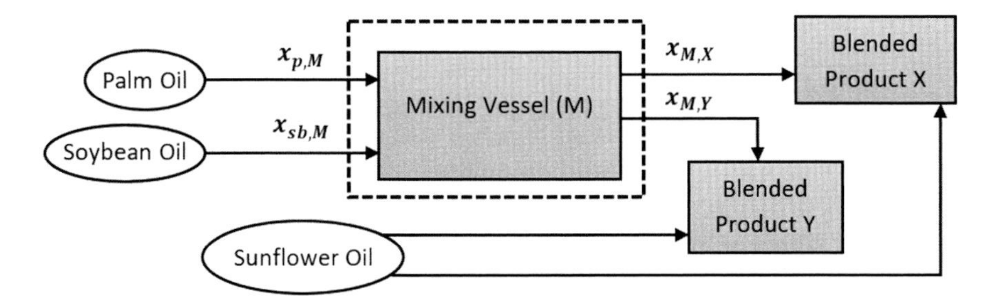


The corresponding mass balance for this control volume can be written as follows, assuming a constant density for all the oils.

Mass of oil into control volume = Mass of oil out of control volume

Assuming all oils have the same density, then

Volume of oil into control volume = Volume of oil out of control volume

$$x_{p,M} + x_{sb,M} = x_{M,X} + x_{M,Y} \cdots \boxed{1}$$

Similarly, we can construct two other mass balance equations by drawing control volumes around the vessel containing product X and vessel containing product Y as shown below. The additional annotations are defined as follows:

- $x_{sf,\ X}$ refers to the number of volume units of sunflower oil entering the vessel containing X.
- $x_{sf,\ Y}$ refers to the number of volume units of sunflower oil entering the mixing vessel containing Y.
- $x_X$ refers to the number of volume units of oil in the vessel containing X.
- $x_Y$ refers to the number of volume units of oil in the vessel containing Y.

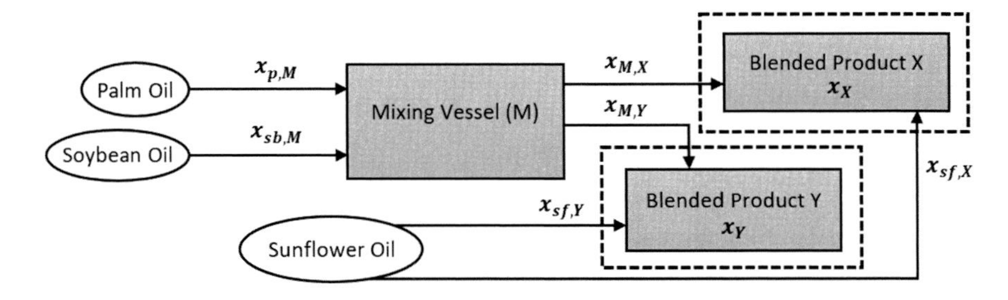


The mass balance equation for the control volume around the vessel containing X can be written as follows

$$x_{M,X} + x_{sf,X} = x_X \cdots \boxed{2}$$

Similarly, the mass balance equation for the control volume around the vessel containing Y can be written as follows

$$x_{M,Y} + x_{sf,Y} = x_Y \cdots \boxed{3}$$

Now that we have three mass balance equations for oil, we can perform another set of mass balances for the saturated fatty acids present in oil. Note that each mass balance equation is specific to a particular (or single) species (e.g. oil, or fatty acids etc.), hence we need to be clear about which species we are balancing.

Going back to the same control volumes, we now perform mass balances for fatty acids, starting with the control volume around the mixing vessel M. The additional annotations are defined as follows:

- 2.5% refers to the percentage of fatty acids in palm oil as given in the problem.
- 0.9% refers to the percentage of fatty acids in soybean oil as given in the problem.
- $y_M$ refers to the percentage of fatty acids within mixing vessel M, which is the same as that in all outlet streams from vessel M, i.e. the stream going into product vessel X (corresponding to $x_{M,\ X}$) and the stream going into product vessel Y (corresponding to $x_{M,\ Y}$).

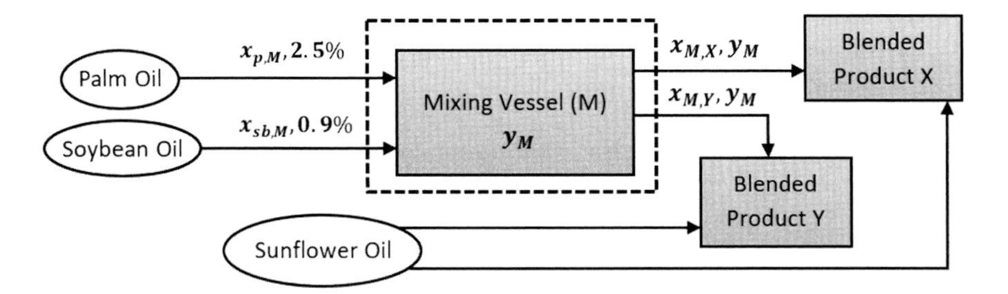


The mass balance for fatty acids for the control volume around mixing vessel M can be written as follows

$$2.5x_{p,M} + 0.9x_{sb,M} = y_M(x_{M,X} + x_{M,Y}) \cdots \boxed{4}$$

Similarly, we can write mass balances for fatty acids, for the two other control volumes around the product vessels containing X and Y respectively. The additional annotations are defined as follows:

- 1% refers to the percentage of fatty acids in sunflower oil as given in the problem. Note that this same percentage should apply for the stream entering product vessel X (corresponding to $x_{sf,\ X}$) and the stream entering product vessel Y (corresponding to $x_{sf,\ Y}$) as shown below.
- $y_X$ and $y_Y$ refer to the percentages of fatty acids in product vessels X and Y respectively.

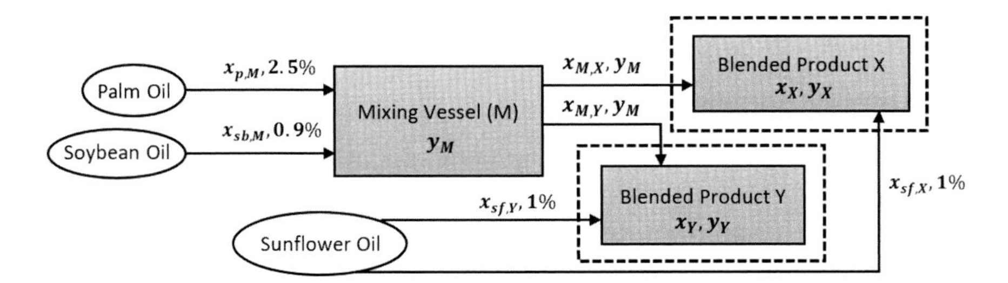


The mass balance for fatty acids for the control volume around product vessel X can be written as follows

$$y_M x_{M,X} + (1)x_{sf,X} = y_X x_X \cdots \boxed{5}$$

The mass balance for fatty acids for the control volume around product vessel Y can be written as follows

$$y_M x_{M,Y} + (1)x_{sf,Y} = y_Y x_Y \cdots \boxed{6}$$

We have now got a total of 6 constraint equations from the mass balances.

We now need to consider if there are any other constraints to be established. Note that it is necessary to fully define our model by listing down all required constraints. Otherwise, a viable solution cannot be obtained (e.g. a lack of constraints leads to non-convergence).

There are quantities in this model that need to be non-negative for them to be physically meaningful. In other words, we will need to introduce non-negativity constraints as shown below:

$$y_X, y_Y, y_M, x_{M,X}, x_{M,Y}, x_{p,M}, x_{sb,M}, x_{sf,X}, x_{sf,Y}, x_X, x_Y \geq 0$$

The next constraint we now consider would be the purchase order limits for the two products X and Y. As stated in the problem, the volume limit for X is 120 units while that for Y is 250 units, therefore,

$$x_X \leq 120$$

$$x_Y \leq 250$$

Another constraint stipulated in the problem (see table) is the percentage of saturated fatty acids in products X ad Y. Product X must not exceed 2% fatty acids, while Y must not exceed 1.2% fatty acids. Therefore we establish two further constraints as shown:

$$y_X \leq 2$$

$$y_Y \leq 1.2$$

Now that we have completely listed down all our constraints, we can write down our objective function as the final step, and optimization of this function by maximization will give us solutions that correspond to maximum profit. If objective function $F$ is written as equivalent to total profit, then

$$F = 80x_X + 140x_Y - \left(50x_{p,M} + 150x_{sb,M}\right) - 100\left(x_{sf,X} + x_{sf,Y}\right)$$

And profit maximization can be done by maximization of $F$ or minimization of $-F$

$$\max F = max\left\{80x_X + 140x_Y - \left(50x_{p,M} + 150x_{sb,M}\right) - 100\left(x_{sf,X} + x_{sf,Y}\right)\right\}, \text{or}$$

$$\min\left(-F\right) = \min\left\{-\left(80x_X + 140x_Y - \left(50x_{p,M} + 150x_{sb,M}\right) - 100\left(x_{sf,X} + x_{sf,Y}\right)\right)\right\}$$

**(b)**

With a convex objective function, we will have a convex feasible region, which means that there can only be one optimal solution, which is globally optimal. However, as this problem is non-convex, there will be multiple feasible regions and hence multiple locally optimal points. The non-convexity arises due to the mixing vessel M which creates bilinear terms involving unknown possible combinations of inlet flows of palm oil and soybean oil, and consequently an unknown fatty acid content of the contents in the vessel M.

**(c)**

By removing the mixing vessel, we would remove non-linearity from the problem. However we should still maintain the integrity of our process setup such that products X and Y can still receive certain amounts of palm oil, soybean oil and sunflower oil. With this in mind, we can modify the flowsheet diagram to the one below, whereby, each of the three oils has a direct inlet to each of the two product vessels (for X and Y).

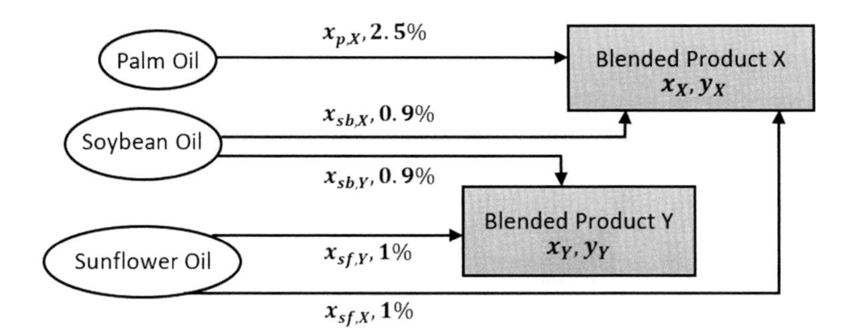


The model can now be reformulated as a linear program as follows:

For a mass balance on oil for the control volume around the vessel containing X, we have

$$x_{p,X} + x_{sb,X} + x_{sf,X} = x_X \cdots \boxed{1}$$

We can write Eq. ($\boxed{1}$) in matrix form as shown below. We can see that we have a system of linear equations written in the matrix form $Ax = B$. Note that the dot product of a $1 \times 3$ and $3 \times 1$ matrix gives a resulting $1 \times 1$ matrix.

$$\begin{pmatrix} 1 & 1 & 1 \end{pmatrix} \begin{pmatrix} x_{p,X} \\ x_{sb,X} \\ x_{sf,X} \end{pmatrix} = \begin{pmatrix} x_X \end{pmatrix}$$

For a mass balance on fatty acid for the control volume around the vessel containing X, we have

$$2.5x_{p,X} + 0.9x_{sb,X} + (1)x_{sf,X} = y_X x_X \cdots \boxed{2}$$

We can write Eq. ($\boxed{2}$) in matrix form as well as shown below. Like before, this matrix equation is also linear as it contains a system of linear equations.

$$\begin{pmatrix} 2.5 & 0.9 & 1 \end{pmatrix} \begin{pmatrix} x_{p,X} \\ x_{sb,X} \\ x_{sf,X} \end{pmatrix} = (y_X x_X)$$

We can now further define our constraint on the fatty acid content for product X, at an upper limit of 2%.

$$0 \le y_X x_X \le 2 \cdots \boxed{3}$$

By solving the set of linear constraints and Eqs. ($\boxed{1}$), ($\boxed{2}$) and ($\boxed{3}$), we will be able to find our solution to the linear model, which would tell us the volumes of each oil present in product X.

Repeating the same steps for product Y, we would arrive at the following equations and constraints, which we can then solve as a linear programming problem, and obtain solutions for the volume composition of oils in product Y.

$$x_{p,Y} + x_{sb,Y} + x_{sf,Y} = x_Y \cdots \boxed{1}$$

$$2.5x_{p,Y} + 0.9x_{sb,Y} + (1)x_{sf,Y} = y_Y x_Y \cdots \boxed{2}$$

$$0 \le y_Y x_Y \le 1.2 \cdots \boxed{3}$$

**(d)**

The solution is viable, because it satisfies all constraints.

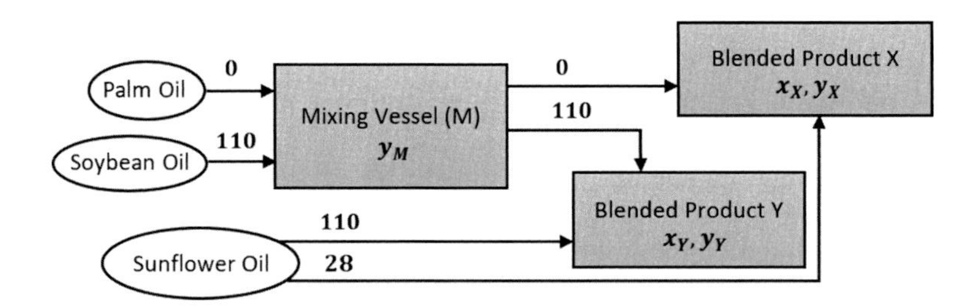


- All volumes are non-negative which satisfy the non-negativity constraint.
- Mass balances are satisfied.

  – $x_{p,M} + x_{sb,M} = x_{M,X} + x_{M,Y}$

    - $0 + 110 = 0 + 110$
    - $110 = 110$

- $x_{M,X} + x_{sf,X} = x_X$
  - $0 + 28 = x_X$
  - $x_X = 28$
- $x_{M,Y} + x_{sf,Y} = x_Y$
  - $110 + 110 = x_Y$
  - $x_Y = 220$

- Requirements on saturated fatty acid content are satisfied. As calculated below, the fatty acid percentage in product X, $y_X = 1$ which satisfies $y_X \leq 2$, while the fatty acid percentage in product Y, $y_Y = 0.95$ which satisfies $y_X \leq 1.2$.
  - $2.5x_{p,M} + 0.9x_{sb,M} = y_M(x_{M,X} + x_{M,Y})$
    - $2.5(0) + 0.9(110) = y_M(0 + 110)$
    - $y_M = 0.9$
  - $y_M x_{M,X} + (1)x_{sf,X} = y_X x_X$
    - $0.9(0) + (1)28 = y_X(28)$
    - $y_X = 1$
  - $y_M x_{M,Y} + (1)x_{sf,Y} = y_Y x_Y$
    - $0.9(110) + (1)110 = y_Y(220)$
    - $y_Y = 0.95$

The value of the objective function can be calculated as shown below, which gives a total profit of $2740 for this particular solution.

$$F = 80x_X + 140x_Y - (50x_{p,M} + 150x_{sb,M}) - 100(x_{sf,X} + x_{sf,Y})$$

$$F = 80(28) + 140(220) - (50(0) + 150(110)) - 100(28 + 110) = 2740$$

Note that in this solution, the mixing vessel is not effectively functioning as a mixer, since there is no palm oil entering it. Only soybean oil enters the mixing vessel and the same volume that flows in then flows out of it, before entering the next vessel containing product Y. Therefore, the mixing vessel (M) simply transfers the same volume of soybean oil to the vessel containing Y.

This solution although proven viable (as it satisfies all constraints) may not be the only solution possible since this is a non-linear problem. Non-convexity due to non-linearity may result in more than one optimization point, i.e. if we minimized $(-F)$, we may obtain other local minima which may provide better solutions for our profit maximization objective.

## Problem 8

**A car makes a journey between two pit stops. The total length of the route is 800 m, and it can be broken down into 3 parts as follows:**

- **Part 1 – From distance 0 to 200 m, the speed limit is 15 m/s**
- **Part 2 – From distance 200 to 450 m, the speed limit is 20 m/s**
- **Part 3 – From distance 450 to 800 m, the speed limit is 25 m/s**

**Given that the car's upper and lower acceleration limits are $-2$ m/s$^2$ and +1.5 m/s$^2$,**

(a) **Formulate an optimization model to minimize total time taken for the journey.**
(b) **Using the backward (implicit) Euler method, which allows different numbers of discrete elements over different periods of the journey, show how this problem can be formulated as a finite dimensional optimization problem, and state any assumptions made.**
(c) **Comment on the source of non-linearity in the discretized optimization problem, and explain whether this reformulated problem is convex or non-convex.**

## Solution 8

**(a)**
We can construct a simple timeline as shown below to better understand the route. $t_3$ refers to the total time taken to travel the full length of the route, while $t_1$ and $t_2$ are the intermediate times at the points when the car has completed the first part of the route (at the 200 m mark) and the second part of the route (at the 450 m mark) respectively.

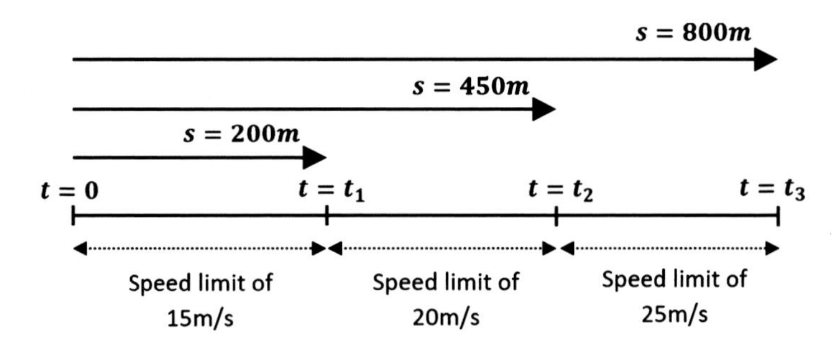


In order to minimize total time taken to travel the full route, we can minimize our objective function $t_3$ as shown below. The optimization parameters (or decision variables) are acceleration $a$, as well as the times $t_1$, $t_2$ and $t_3$ as shown below the word "min".

$$\min_{a, t_1, t_2, t_3} t_3$$

This model is not complete yet, as we need to establish all constraints to fully define our problem.

We know that acceleration $a$ is defined as the rate of change of velocity $v$, and $v$ is defined as the rate of change of displacement $s$, therefore we can establish these kinematics relationships (Newton's equations of motions) as follows:

$$\frac{ds}{dt} = \dot{s} = v(t)$$

$$\frac{dv}{dt} = \dot{v} = a(t)$$

Constraints on speed limit:

$$v(t) \leq 15, \quad t \in (0, t_1)$$

$$v(t) \leq 20, \quad t \in (t_1, t_2)$$

$$v(t) \leq 25, \quad t \in (t_2, t_3)$$

Constraints on car's acceleration:

$$-2 \leq a \leq 1.5$$

Initial conditions at $t = 0$ are fixed as follows, since the car has to start from rest at the zero displacement position:

$$s(0) = 0$$

$$v(0) = 0$$

Final conditions for each part of the journey are also fixed as follows. Note that in part 3, there is an additional condition for velocity because the car has to come to a stop when the journey is completed:

<u>Part 1:</u>

$$s(t_1) = 200$$

Part 2:

$$s(t_2) = 450$$

Part 3:

$$s(t_3) = 800$$
$$v(t_3) = 0$$

With all the above constraints set, the objective function can be minimized to determine the optimization point that corresponds to the shortest travel time required for the journey.

**(b)**
Using the discretization method, we can break up time into a series of small elements or "time steps". The cumulative addition of all these small steps within a defined interval will then give the total time over that interval. The smaller the step size, the more accurate the method will be in approximating to the actual time within the interval.

We can first define the size of one time step as $h$, and the index used to count the steps can be denoted $i$. We then establish equations that correlate step size $h$ at each step, with the total time elapsed by summing all steps within a defined interval.

These equations serve as "instructions" in our model that help guide the program in handling the new variable $h$ in the optimization process.

- For part 1 of the journey, assuming we take $(N_1 - 1)$ number of time steps, for an interval defined from $t = 0$ to $t = t_1$. Summing all steps within this time interval will therefore give us the total time elapsed which is $t_1$ or the time taken for part 1 of the journey.

$$t_1 = \sum_{i=1}^{N_1} h^{(i)} = h^{(1)} + h^{(2)} \ldots + h^{(N_1)}$$

Referring to the summation notation above, the first step occurs at $i = 1$, and the second step at $i = 2$ and so on. When we sum all the time steps until the last step at $i = N_1$, we would have taken a total of $N_1$ steps in the time from $t = 0$ to $t = t_1$.

Note that we are assuming that the step sizes are unequal free variables, hence each step may not be exactly the same size, and they are differentiated by the superscript, i.e. $h^{(1)}$ is the size of step 1, $h^{(2)}$ is the size of step 2 and so on, and $h^{(1)}$ need not be equal to $h^{(2)}$.

A diagram showing part 1 of the journey is shown below:

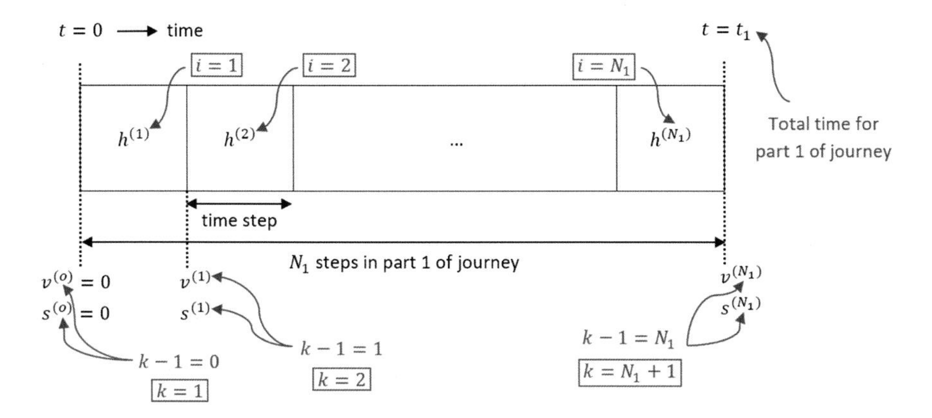


For considerations of motion, e.g. displacement and velocity, we will define a counting index $k$, whereby:

- At $t = 0$, $s = s^{(0)}$, $v = v^{(0)}$, $a = a^{(0)}$, and this corresponds to $k - 1 = 0$ or $k = 1$.
- At $t = 0 + h^{(1)}$, $s = s^{(1)}$, $v = v^{(1)}$, $a = a^{(1)}$, and correspond to when $k - 1 = 1$ or $k = 2$.
- At $t = 0 + h^{(1)} + h^{(2)}$, $s = s^{(2)}$, $v = v^{(2)}$, $a = a^{(2)}$, and correspond to when $k - 1 = 2$ or $k = 3$.
- Finally at the end of part 1 of the journey, we have $t_1 = \sum_{i=1}^{N_1} h^{(i)} = 0 + h^{(1)} + h^{(2)} \ldots + h^{(N_1)}$, $s = s^{(N_1)}$, $v = v^{(N_1)}$, $a = a^{(N_1)}$, and correspond to when $k - 1 = N_1$ or $k = N_1 + 1$

- For part 2 of the journey, we similarly assume a particular number of steps, e.g. $N_2$ steps from $t = t_1$ to $t = t_2$. Summing over these $N_2$ steps will give us the time taken for part 2 of the journey which is $t_2 - t_1$.

$$t_2 - t_1 = \sum_{i=N_1+1}^{N_1+N_2} h^{(i)} = h^{(N_1+1)} + h^{(N_1+2)} \ldots + h^{(N_1+N_2)}$$

Alternatively we can express the total time from $t = 0$ to $t = t_2$, which includes parts 1 and 2 of the journey as follows:

$$t_2 = \sum_{i=1}^{N_1+N_2} h^{(i)} = h^{(1)} + h^{(2)} \ldots + h^{(N_1+N_2)}$$

A diagram showing part 2 of the journey is shown below:

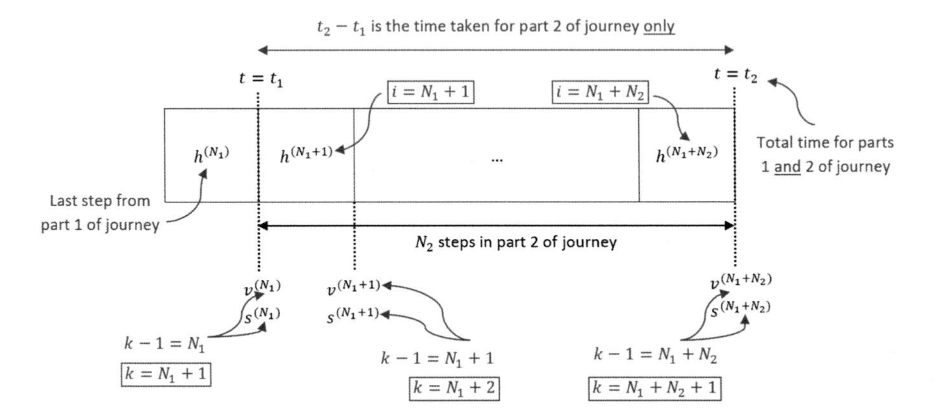


In part 2 of the journey, the considerations of motion are as follows:

- We have to take $N_2$ steps for part 2. From $t = t_1$ to $t = t_1 + h^{(N_1+1)}$ is our first step in part 2. After taking $N_2$ steps, we will arrive at $t = t_2$ which occurs right after we have taken the last step in part 2 which is $h^{(N_1+N_2)}$ and this point corresponds to $s = s^{(N_1+N_2)}$, $v = v^{(N_1+N_2)}$, $a = a^{(N_1+N_2)}$ and $k - 1 = N_1 + N_2$ or $k = N_1 + N_2 + 1$.

- For part 3 of the journey, we assume $N_3$ number of time steps from $t = t_2$ to $t = t_3$. Summing over these $N_3$ steps will give us the time taken for part 3 of the journey which is $t_3 - t_2$.

$$t_3 - t_2 = \sum_{i=N_1+N_2+1}^{N_1+N_2+N_3} h^{(i)} = h^{(N_1+N_2+1)} + h^{(N_1+N_2+2)} \ldots + h^{(N_1+N_2+N_3)}$$

Alternatively we can express the total time from $t = 0$ to $t = t_3$ (also the total time for the entire journey), which includes parts 1, 2 and 3 of the journey as follows.

$$t_3 = \sum_{i=1}^{N_1+N_2+N_3} h^{(i)} = h^{(1)} + h^{(2)} \ldots + h^{(N_1+N_2+N_3)}$$

A diagram showing part 3 of the journey is shown below:

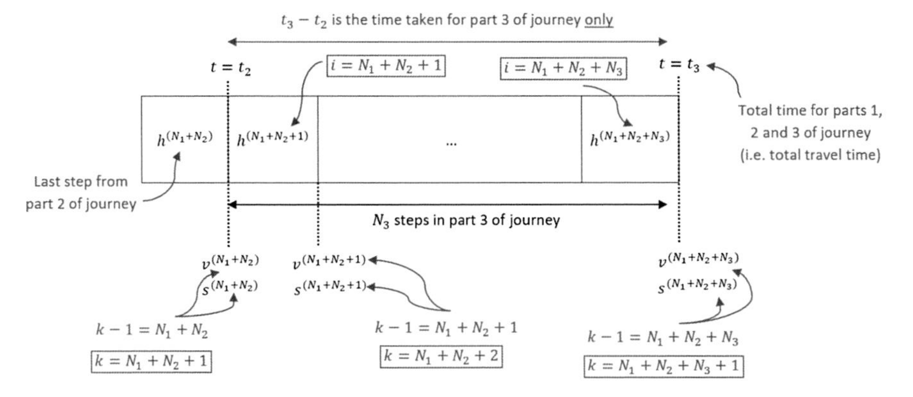


In part 3 of the journey, the considerations of motion are as follows:

– We have to take $N_3$ steps for part 3. From $t = t_2$ to $t = t_2 + h^{(N_1+N_2+1)}$ is our first step in part 3. After taking $N_3$ steps, we will arrive at $t = t_3$ which occurs right after we have taken the last step in part 3 which is $h^{(N_1+N_2+N_3)}$ and this point corresponds to $s = s^{(N_1+N_2+N_3)}$, $v = v^{(N_1+N_2+N_3)}$, $a = a^{(N_1+N_2+N_3)}$ and $k - 1 = N_1 + N_2 + N_3$ or $k = N_1 + N_2 + \ + N_3 + 1$.

- We can see from the above that $i$ and $k$ takes values as follows.

$$i = 1, 2, 3 \ldots N_1 + N_2 + N_3$$

- Observe how Euler's method uses the gradient near to the point of interest to predict the next iteration value. For the implicit method particularly, we need to write an equation in terms of $k - 1$, hence the minimum value of $k$ is 1 (and not 0). The range of values $k$ can take are as follows.

$$k = 1, 2, 3 \ldots N_1 + N_2 + N_3 + 1 \cdots \boxed{1}$$

***How does Euler's backward method work?***

In the case of displacement $s$, the value of $s$ in the $k^{th}$ step or $s^{(k)}$ can be estimated from the $s$ value in the previous $(k - 1)^{th}$ step or $s^{(k-1)}$, by adding an incremental value $\Delta s$ to it. This translates to

$$s^{(k)} = s^{(k-1)} + \Delta s \cdots \boxed{2}$$

This is shown in the diagram below.

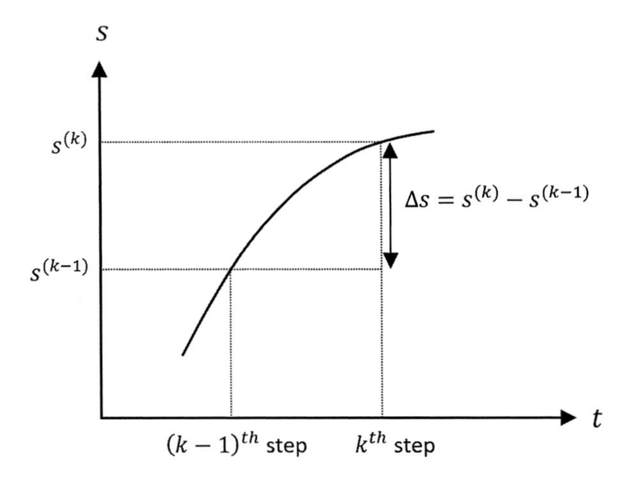


For a displacement-time graph, where $h$ is a time step along the horizontal time axis, the gradient of a straight line connecting $s^{(k-1)}$ to $s^{(k)}$ is also equal to the rate of change of displacement $s$ over the time step of size $h^{(k)}$. Note that the smaller the step size, the more accurate this straight line gradient approximation method will be.

$$Gradient = \frac{Rise}{Run} = \frac{s^{(k)} - s^{(k-1)}}{time\ step} = \frac{\Delta s}{h^{(k)}}$$

We know that the rate of change of displacement with time is simply velocity, hence

$$Gradient = \frac{\Delta s}{h^{(k)}} = v^{(k)} \ldots \boxed{3}$$

Substituting Eq. ($\boxed{3}$) back into Eq. ($\boxed{2}$)

$$s^{(k)} = s^{(k-1)} + h^{(k)}v^{(k)}$$

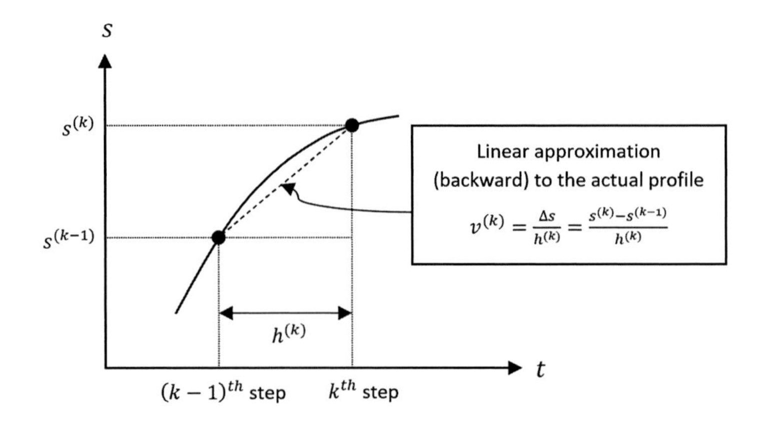

We can now follow the same process for velocity, and this would give us a similar equation as shown below, where acceleration $a$ is the rate of change of velocity with time. Note that the expression below mimics the form of the equation we got for displacement earlier, whereby velocity was the rate of change of displacement with time.

$$v^{(k)} = v^{(k-1)} + h^{(k)} a^{(k)}$$

Boundary conditions

We can next define our initial conditions at $t = 0$, which is when the car is at rest and has not moved any displacement, hence there is zero velocity and displacement at the initial position that corresponds to when $k - 1 = 0$ or $k = 1$ (the lowest value in its range, refer to Eq. (1) above):

$$s^{(0)} = 0$$

$$v^{(0)} = 0$$

Final conditions for each part of the journey can also be defined as follows:

Part 1:

$$s_1|_{t=t_1} = s^{(N_1)} = 200$$

Part 2:

$$s_2|_{t=t_2} = s^{(N_1+N_2)} = 450$$

Part 3:

$$s_3|_{t=t_3} = s^{(N_1+N_2+N_3)} = 800$$

$$v|_{t=t_3} = v^{(N_1+N_2+N_3)} = 0$$

Other constraints

And next our constraints on speed limit can be written:

$$v^{(k)} \leq 15, \;\; k = 1, 2 \ldots N_1 + 1$$

$$v^{(k)} \leq 20, \;\; k = N_1 + 2, N_1 + 3 \ldots N_1 + N_2 + 1$$

$$v^{(k)} \leq 25, \;\; k = N_1 + N_2 + 2, N_1 + N_2 + 3 \ldots N_1 + N_2 + N_3 + 1$$

And the constraint on the car's acceleration can be expressed as:

$$-2 \leq a^{(k)} \leq 1.5$$

Finally we need to establish any non-negativity constraints, which would apply for the step size.

$$h^{(i)} \geq 0, \;\; i = 1, 2, 3 \ldots N_1 + N_2 + N_3$$

The above equations now fully define our discretized model using the backwards Euler's method. Let us now restate our objective function such that we include all optimization parameters in the reformulated discretized model as shown below.

$$\min_{h^{(i)}, a^{(k)}, s^{(k)}, v^{(k)}, t_1, t_2, t_3} \quad t_3$$

**(c)**

The discretized problem in part b is non-linear due to the presence of bilinear terms. Linearity occurs when the objective function and constraints consist only of linear equalities and/or inequalities. Non-linearity arises when at least one of the objective functions and constraints is non-linear. Bilinear terms are a source of non-linearity in optimization.

In part b, the reformulated discretized problem had introduced unknown variables such as step size $h$ which forms bilinear terms with the problem's original variables such as $v$ and $u$. We can further deduce that the reformulated optimization problem in part b is non-convex due to the bilinear terms. A non-convex problem results in multiple feasible regions, and hence multiple local optima. This is as opposed to a convex problem, where there is only one optimal solution in a convex feasible region, or the global optima.

## Problem 9

**A delivery man was tasked to fit three cylindrical containers of the same size into a rectangular box. The radius of the circular flat surface of each cylinder is *R* while their axial lengths are denoted *L*. The base of the rectangular box has a length *Q* and a width *P*. The delivery man arranged the cylinders such that their flat surfaces face down, with their axial lengths at right angles to the laid surface as shown below.**

**Top view:**

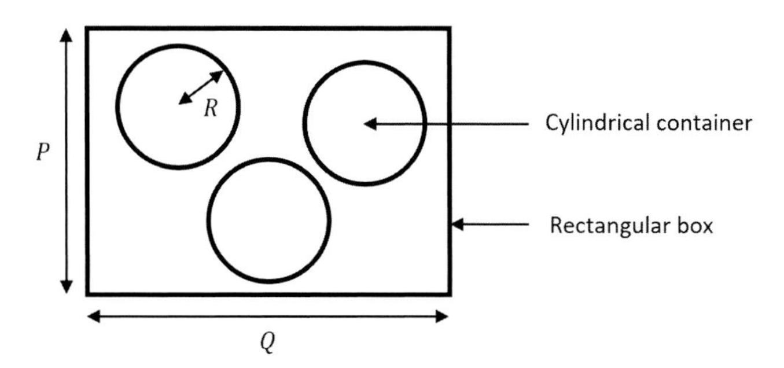


**Side view:**

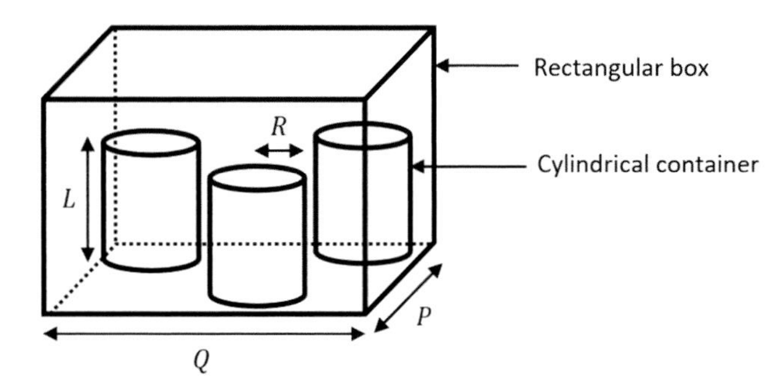


(a) **Formulate a model for this optimization problem such that the perimeter of the box is minimized. You may use coordinates $x_i, y_i$ whereby $i = 1,2$ or 3 for the three cylinders to mark the center of the circular faces of the cylinders in contact with the base of the box, and assume that $x$ and $y$ represent the coordinate directions along the length $Q$ and width $P$ of the box respectively. Comment on the type of optimization problem this is.**

(b) **Express your model in part a in a general form for an arbitrary number of $N$ cylinders and determine the number of constraints this generalized optimization model would have.**

(c) **When the optimization model was applied by the deliveryman for multiple cylinders, he realized that different solutions were obtained depending on the starting point chosen. Explain this behavior in terms of the likely cause, and the part of the model where it is located. Is it possible to reformulate the model to avoid this behavior?**

## Solution 9

**(a)**
In this problem, the objective function is the perimeter of the rectangular base of the box, and the required optimization would be a minimization of this perimeter. Therefore we can express this minimization as follows:

Minimization of objective function:

$$\min\left[Perimeter\right] = \min\left[2(P+Q)\right]$$

Since the constant term 2 does not affect the optimization, it can be taken out of the objective function for simplification.

$$\min\left[2(P+Q)\right] \rightarrow \min\left[P+Q\right]$$

Constraints for all cylinders to fit into box:
In order for all three cylinders to fit within the perimeter of the box, we require the following 4 constraints for each cylinder, which gives us a **total of 12 constraints** for the 3 cylinders.

$$x_i \geq R \cdots \boxed{1}, \quad for\ i = 1,2,3$$

$$y_i \geq R \cdots \boxed{2}, \quad for\ i = 1,2,3$$

$$x_i + R \leq Q \cdots \boxed{3}, \quad for\ i = 1,2,3$$

$$y_i + R \leq P \cdots \boxed{4}, \quad for\ i = 1,2,3$$

Let us examine further how constraints ($\boxed{1}$) and ($\boxed{2}$) came about with the diagram below:

***Moving cylinder away from the corner***
Consider a single cylinder (could be any one of the three) placed right against the bottom left corner of the box with no gaps, as shown below.

Top view:

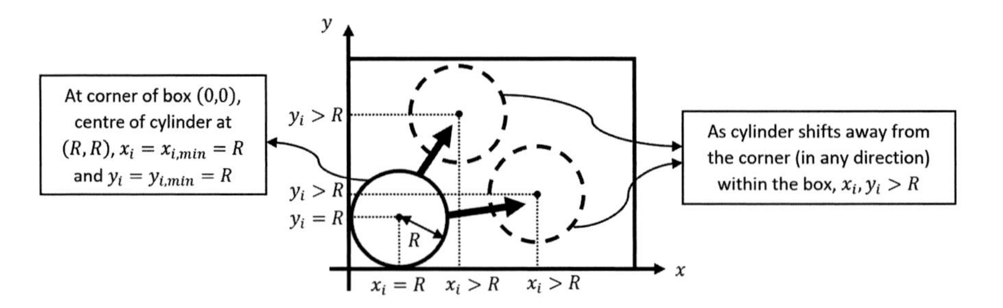


- The corner position marks the minimum limits of $x_i$ and $y_i$ in order for the cylinder to stay within the box. The corresponding coordinates of the center point of this cylinder's face must then be $x_i = R$ and $y_i = R$, whereby $R$ is the minimum value for both coordinates.

- If the cylinder was then shifted further away from this corner, we observe that $x_i$ and $y_i$ start to increase to values greater than $R$. Combining these two observations, we arrive at constraints ([1]) and ([2]), which state that $x_i \geq R$ and $y_i \geq R$.

Let us now examine how constraints ([3]) and ([4]) came about with reference to the two diagrams below:

***Moving cylinder in the x-direction***

We shall again consider a single cylinder, and move it from the bottom left corner horizontally to the right until it just touches the opposite edge of the box, at $x_i = Q - R$. This position corresponds to the maximum value of $x_i$ for the cylinder to remain in the box, therefore $x_i = x_{i,\ max} = Q - R$.

Top view:

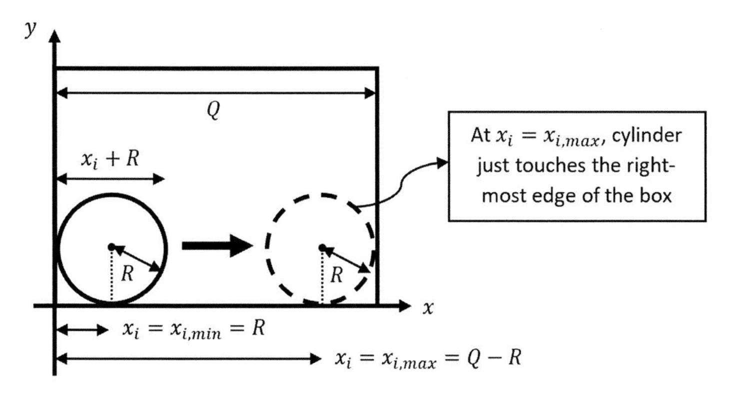


***Moving cylinder in the y-direction***

Likewise, if we moved the cylinder from the bottom left corner vertically upwards until it just touches the top edge of the box, then $y_i = y_{i,\ max} = P - R$. These give rise to our constraints ([3]) and ([4]) obtained earlier.

Top view:

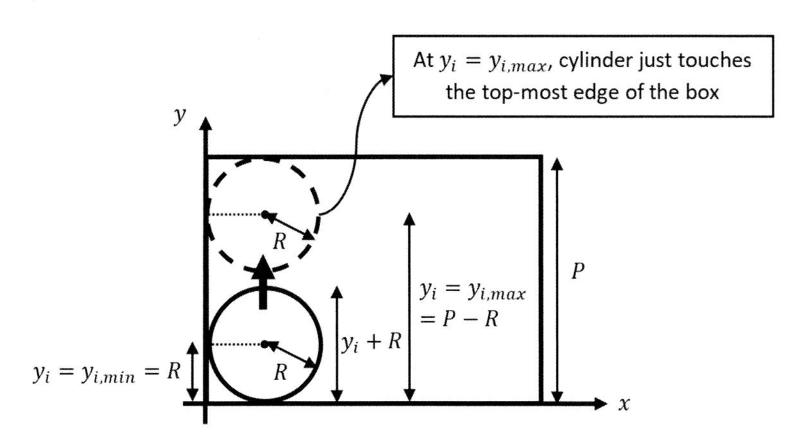

Constraints for non-overlap of cylinders:

Now let us consider another necessary constraint for packing the three cylinders in the box. We should establish rules in our model that ensure the cylinders do not overlap each other. Let us consider a pair of cylinders at a time, for e.g. $i = 1$ and 2. Possible positions of cylinder 2 to 1 such that they are closest together (just touching) without overlap are shown in the diagram as $2a$, $2b$ and $2c$. For positions $2a$, $2b$ and $2c$, we can determine expressions for this closest distance between the centers of the two cylinders.

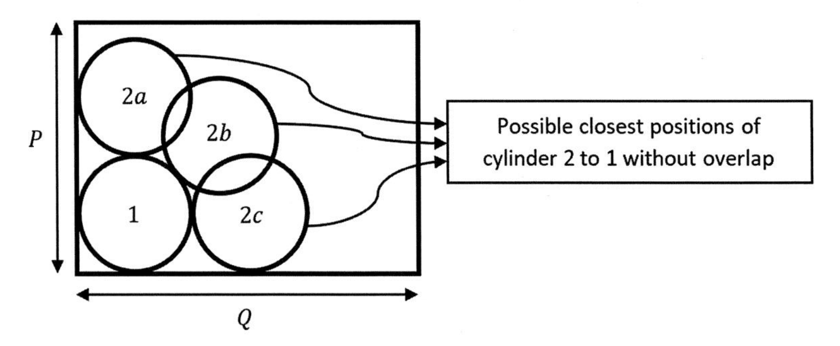


For position $2a$, the distance between centers of cylinders 1 and $2 = y_2 - y_1 = 2R$ as shown in the diagram below.

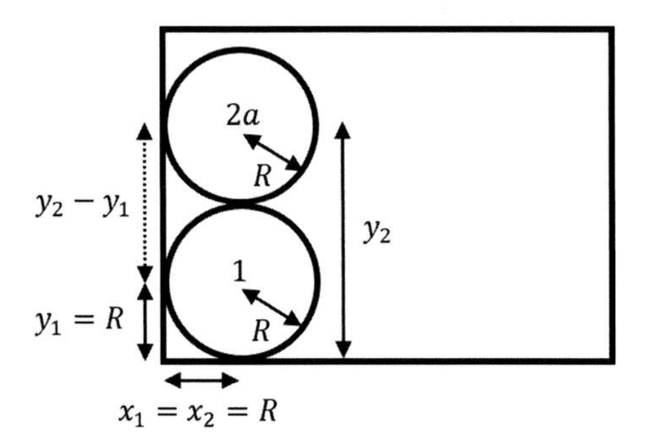


For position $2c$, the distance between centers of cylinders 1 and $2 = x_2 - x_1 = 2R$ as shown in the diagram below.

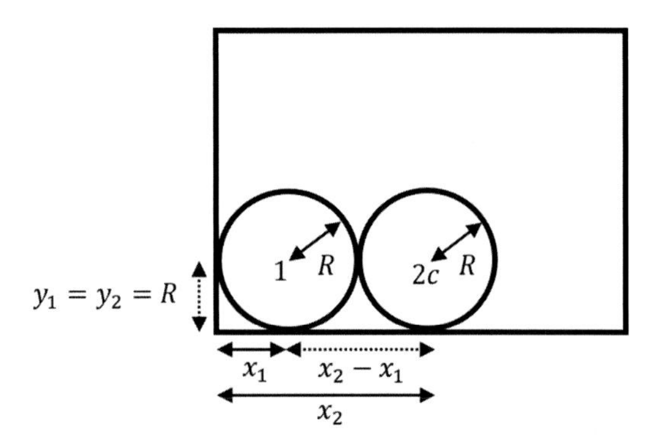

For position $2b$, the distance between centers of cylinders 1 and 2 is equivalent to $2R$, as well as the hypothenuse of a right-angled triangle (marked in red in the diagram below). Hence we can use Pythagoras' theorem to establish a relationship between $x_1$, $x_2$, $y_1$, $y_2$ and $R$ as shown in the diagram below.

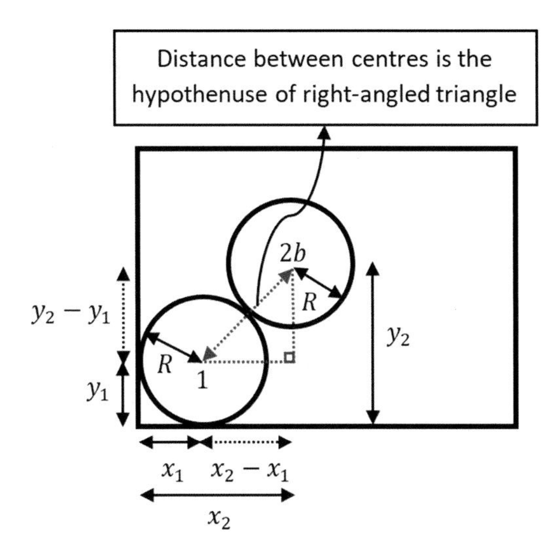


Minimum distance between centers of cylinders 1 and 2 $=2R=$ hypothenuse of right-angled triangle

$$\text{By Pythagoras' theorem}: (2R)^2 = (x_2 - x_1)^2 + (y_2 - y_1)^2$$

Taking only the physically meaningful positive solution of the square root, we have the general form of the expression for the minimum distance as shown below:

$$2R = +\sqrt{(x_2 - x_1)^2 + (y_2 - y_1)^2} \cdots \boxed{*}$$

In fact, we observe that the other two possible positions of closest approach between cylinders 1 and 2, i.e. for cases of $2a$ and $2c$, they can be derived from the general form of the expression ($\boxed{*}$), whereby case $2a$ corresponds to when $x_1 = x_2$ which makes the first term zero (giving us $y_2 - y_1 = 2R$) and for case $2c$, $y_1 = y_2$ which makes the second term zero (giving us $x_2 - x_1 = 2R$).

Since the derived expression ($\boxed{*}$) defines the minimum distance apart, our constraint for non-overlap can be written as an inequality that shows this minimum limit as follows:

$$+\sqrt{(x_2 - x_1)^2 + (y_2 - y_1)^2} \geq 2R$$

We can square both sides of the inequality without changing the direction of the inequality sign (since we are only taking the positive square root on the left hand side). This avoids the complication of dealing with square roots in our model, hence simplifying the expression to the following:

$$(x_2 - x_1)^2 + (y_2 - y_1)^2 \geq (2R)^2$$

We can also swap the orders of $x_1$ and $x_2$, as well as $y_1$ and $y_2$ since the squaring of the terms in brackets will ensure the same result is obtained with this swap.

$$(x_1 - x_2)^2 + (y_1 - y_2)^2 \geq (2R)^2$$

We will now extend the expression for non-overlap to all other possible pairs of cylinders amongst the three cylinders. This gives us three constraint inequalities in total, for the pairs 1-2, 1-3 and 2-3 as shown below.

$$(x_1 - x_2)^2 + (y_1 - y_2)^2 \geq (2R)^2 \cdots \boxed{5}$$

$$(x_1 - x_3)^2 + (y_1 - y_3)^2 \geq (2R)^2 \cdots \boxed{6}$$

$$(x_2 - x_3)^2 + (y_2 - y_3)^2 \geq (2R)^2 \cdots \boxed{7}$$

Constraints on non-negativity of physical quantities:
For physically meaningful values to be obtained from our model, it is good practice to always do a sense-check on the non-negativity constraint. This means that physical quantities such as length, volume, mass etc. should only be non-negative numbers. As such, we add the following constraint to our model to ensure our program is fully defined and able to produce meaningful solution(s).

$$P \geq 0, \quad Q \geq 0 \cdots \boxed{8}$$

Summing up our model formulation:
We can now conclude our optimization model as a minimization of the objective function subject to total of **17 constraint equations / inequalities**. (= 12 from ($\boxed{1}$) to ($\boxed{4}$); 3 from ($\boxed{5}$) to ($\boxed{7}$); 2 from ($\boxed{8}$)) constraints as shown:

$$\min\,[P + Q], \quad \text{s.t.}(1), \text{to } (8)$$

The problem further asks about the type of optimization problem in this case. Let us first explain some fundamentals on the types of optimization problems.

Linear vs non-linear programming problem:
A problem is classified as a non-linear programming problem (NLP) if the objective function is non-linear and/or the feasible region is determined by non-linear constraints (can be equalities or inequalities). On the other hand, we have a linear

programming problem if the converse is true, i.e. the objective function and all constraints are linear.

In part a, our formulated problem introduced non-linear inequality constraints from ($\boxed{5}$) to ($\boxed{7}$) for the non-overlap condition. This confirms our problem is an NLP problem.

Types of NLP problems:

After understanding what NLP problems are, we note that they can be further classified into:

- Strictly convex – Minimization gives a global optimum.
- Strictly concave – Maximization gives a global optimum.
- Non-convex and non-concave – Multiple local optima (maxima and/or minima) are possible.

The overlap constraints that caused our problem to be non-linear, are also non-convex functions. Therefore, multiple solutions (corresponding to multiple local optima) can be expected for this optimization problem.

Examples of strictly convex and strictly concave functions

Whenever a function curves "upwards" as shown below left, it is strictly convex, and the minimum point is also the global (i.e. no other minimum points present) minimum. Conversely, if a function curves "downwards" as shown below right, it is strictly concave, and the maximum point is also the global maximum.

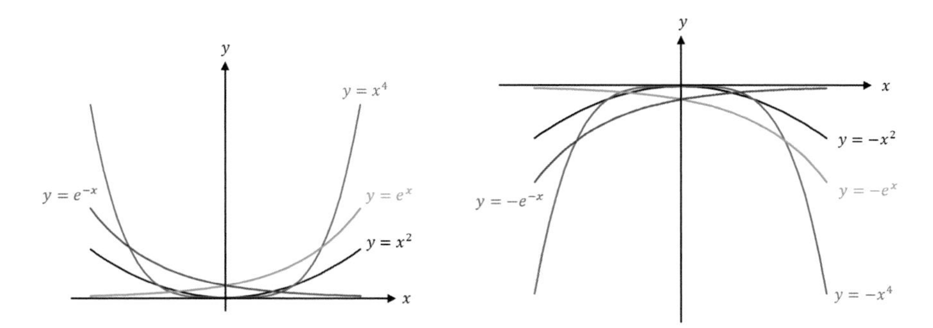


We note that we can easily convert a strictly convex function into a strictly concave function by simply adding a negative sign. This makes sense since the directionality of the function reverses to its exact opposite with negation. It is useful to note that we can also obtain convex functions by summing up multiple convex functions, or by having a non-negative multiple of a convex function. This is the same for concave functions.

Going back to our problem, we notice that the non-overlap constraints ($\boxed{5}$) to ($\boxed{7}$) do not have the form of expression required to be a strictly convex function for minimization. As such, they are non-convex.

$$(x_1 - x_2)^2 + (y_1 - y_2)^2 \geq (2R)^2 \cdots \boxed{5}$$

$$(x_1 - x_3)^2 + (y_1 - y_3)^2 \geq (2R)^2 \cdots \boxed{6}$$

$$(x_2 - x_3)^2 + (y_2 - y_3)^2 \geq (2R)^2 \cdots \boxed{7}$$

**(b)**

For an arbitrary number of *N* cylinders, we would still have the same minimization of perimeter as in part a, therefore, the following remains unchanged.

Minimization of objective function:

$$\min\,[P + Q]$$

Constraints for all cylinders to fit into box:

As for the earlier constraints ($\boxed{1}$) to ($\boxed{4}$) in part a which ensure all cylinders fit into the box, the only change would be to have values for *i* going from 1 to *N* for *N* cylinders, instead of the earlier 1 to 3 for 3 cylinders. This gives us a total of 4*N* constraints for the *N* cylinders.

$$x_i \geq R \cdots \boxed{1}, \quad for\ i = 1, 2, \dots N$$

$$y_i \geq R \cdots \boxed{2}, \quad for\ i = 1, 2, \dots N$$

$$x_i + R \leq Q \cdots \boxed{3}, \quad for\ i = 1, 2, \dots N$$

$$y_i + R \leq P \cdots \boxed{4}, \quad for\ i = 1, 2, \dots N$$

Constraints for non-overlap of cylinders:

Next, we look at the constraints for no overlap between cylinders. Earlier for 3 cylinders, we had considered 3 possible pairs which were 1-2, 1-3 and 2-3. Now that we have *N* cylinders, there would be ${}^{N}C_2$ number of ways to select distinct pairs from the set of *N* cylinders. This gives us ${}^{N}C_2$ number of constraints for the non-overlap condition.

Recall that the formula for the general form ${}^{N}C_r$, also referred to as "*N* choose *r*" under the topic of 'Permutation & Combination' in mathematics,

$$r_r = \frac{N!}{r!(N - r)!}$$

In this problem, the number of ways to choose a distinct pair of cylinders (i.e. 2) from *N* cylinders can be expressed as:

$$\mathrm{CN}_2 = \frac{N(N - 1)}{2}$$

Therefore, we now have pairs such as 1-2, 1-3, 1-4. . .1-*N*, 2-3, 2-4, 2-5. . .2-*N*, 3-4, 3-5, 3-6. . .3-*N* and so on. And the way we can express this non-overlap constraint in a form that includes all possible distinct pairs (without duplication) is as follows, where the range of values for $i$ and $j$ need to be specified:

$$(x_i - x_j)^2 + (y_i - y_j)^2 \geq (2R)^2 \cdots \boxed{5}, \quad i = 1, 2, \ldots, (N-1), \quad j = i+1, i+2, \ldots, N$$

Note that like before in part a, the non-overlap inequality constraints are non-linear, hence the problem in part b is still an NLP problem. In addition to being non-linear, the non-overlap constraints are also non-convex, like in part a.

Constraints on non-negativity of physical quantities:
Finally we include our non-negativity constraint which still holds true for $N$ cylinders, as follows.

$$P \geq 0, \quad Q \geq 0 \cdots \boxed{6}$$

Summing up our model formulation:
We can now conclude our generalized optimization model for $N$ cylinders as a minimization of the objective function subject to total of $\mathbf{4N + \frac{N(N-1)}{2} + 2}$ **constraint equations/inequalities**. (= $4N$ from ($\boxed{1}$) to ($\boxed{4}$); $\frac{N(N-1)}{2}$ from ($\boxed{5}$); 2 from ($\boxed{6}$)) constraints as shown:

$$\min[P + Q], \quad s.t. \boxed{1} \; to \; \boxed{6}$$

**(c)**
The problem can produce multiple solutions due to the non-convexity of the **non-overlap constraints** established in the model formulation. Only for a strictly convex, or strictly concave function, the mode will converge to a single unique solution that corresponds to the global optimum (one could also say that the local optimum is equivalent to the global optimum, of which there is one).

In cases of non-convexity, multiple solutions may arise due to the presence of multiple local optimas which are each their respective optimas in their respective local feasible regions. However, these local optimas are not global and unique, since there are several of them in such non-convex functions.

It is not possible to reformulate the model in this problem to avoid the issue of multiple solutions, as the problem is already fully specified through the stipulated objective function, optimization variables and constraints. Within the confines of the information given, the formulated model contains no redundant or insufficient information, and cannot be further simplified. Furthermore, the non-convexity of non-overlap constraint arises due to the nature of the case scenario, and there is no straightforward mathematical manipulation to convert non-convex functions into convex forms simply by reformulation due to their marked differences.

## Problem 10

**The mechanism for a particular chemical reaction carried out in a batch reactor is shown below, whereby $k$ refers to the reaction rate constant.**

$$P \overset{k_1}{\rightarrow} Q \overset{k_2}{\rightarrow} R$$

**The rate constants can be correlated to temperature via the Arrhenius equation as shown below:**

$$k_i = A_i \, exp\,(-E_i/RT), \quad i = 1, 2$$

**It is given that both reactions are first-order, and take place under conditions of constant temperature and density. While the reactor temperature is kept constant throughout the reaction at $T$, $T$ can only take values within an allowable range of $T^L \leq T \leq T^U$, between an upper and lower bound. At time $t = 0$, the reaction is charged with pure reactant $P$ at an initial concentration denoted $C_{P0}$. The reactor operates until a final time, $t_f$ and this value is capped at an upper bound of $t_f^U$. At the end of the reactor's operation, it is required that no more than 0.95% of the original amount of $P$ remains in the reactor vessel.**

(a) **Formulate an optimal control problem for the maximization of the concentration of $Q$ at the end of the reactor operation.**
(b) **Sketch the concentration-time profiles of $P$, $Q$ and $R$, indicating the optimum point for the case where $k_1 \gg k_2$, and for the case where $k_1 \cong k_2$. Comment on the significance of any difference between the optimas for the two cases.**
(c) **Derive an expression for the minimum possible value of $t_f^U$ for the model to produce a solution. Describe how one might be able to find the maximum value of $t_f^U$ and state any assumptions made.**

**In most numerical solutions to problems, one common approach is to adopt a discretization scheme.**

(d) **Using a suitable diagram, highlight the key features of the forward and backward difference discretization schemes.**
(e) **In the numerical solution to this problem, an unequal element backward difference discretization scheme was used. Given that $N$ time steps were taken, and that step size was subject to a lower bound of $h_i \geq h^L$, (where $i = 1,2, \ldots N$ identifies the particular elements in the discretization, and $h_i$ refers to the size of a particular element or step), reformulate the optimal control problem in part a as a standard optimization problem using this discretization scheme.**

# Solution 10

**(a)**
We are given the following sequential reactions, with first order rate constants $k_1$ and $k_2$.

$$P \xrightarrow{k_1} Q \xrightarrow{k_2} R$$

From prior knowledge of reaction kinetics and rate laws, we know that we can express the rate of change of species concentrations with time, in terms of rate constants and species concentrations. This allows us to write down the following 3 rate equations for the 3 species involved in reaction.

$$\text{Rate of change of reactant } P = \frac{dC_p}{dt} = -k_1 C_p$$

Note that the rate of change of $P$, $\frac{dC_p}{dt}$ is negative since it is a reactant and will be reacted away with time, hence its concentration $C_p$ decreases with time, and $\frac{dC_p}{dt} < 0$. This gives us our first equation as shown below.

$$\frac{dC_p}{dt} = -k_1 C_p \cdots \boxed{1}$$

As for intermediate $Q$, its overall rate of change (or formation) is equivalent to its rate of formation from reaction 1, minus the rate of its removal by reaction 2. We can express this as follows.

$$\frac{dC_Q}{dt} = k_1 C_p - k_2 C_Q \cdots \boxed{2}$$

Finally the overall rate of change of final product $R$ can be expressed as follows.

$$\frac{dC_R}{dt} = k_2 C_Q \cdots \boxed{3}$$

Differential Eqs. ($\boxed{1}$), ($\boxed{2}$) and ($\boxed{3}$) are to be subject to the following constraints, initial conditions, and boundary conditions.

The time variable $t$ can only take non-negative values, and must also not exceed the total time taken (i.e. final time $t_f$) for the reactor operation. Hence, we have

$$0 \le t \le t_f \cdots \boxed{4}$$

We can define our initial conditions by denoting the initial concentration of $P$ as $C_{P0}$, while the initial concentrations of $Q$ and $R$ and $t = 0$ should both be zero. Hence, we have

$$C_P(t=0) = C_{P0} \cdots \boxed{5}$$

$$C_Q(t=0) = C_{Q0} = 0 \cdots \boxed{6}$$

$$C_R(t=0) = C_{R0} = 0 \cdots \boxed{7}$$

We will now establish the end-point condition that decides when our reactor operations should stop, hence this condition is also a constraint that determines the value of $t_f$. We know that the reactions will be stopped when a composition of 0.95% (of the original amount) of $P$ is reached in the reactor vessel. Hence we have

$$C_P(t=t_f) \le 0.0095 C_{P0} \cdots \boxed{8}$$

Next, we need to remember to explicitly define in our model, the two Arrhenius equations for both reactions, so that the model knows how to correlate rate constant to temperature. Note that the optimization is done with respect to variables $t_f$ and temperature $T$. Therefore, the model needs to know how $T$ relates to the rest of the formulated model.

Rate constants can be correlated to temperature according to the Arrhenius equation as shown below, where $A_i$ refers to a pre-exponential factor, $E_i$ refers to activation energy, $R$ refers to the universal gas constant, and $T$ refers to absolute temperature measured in Kelvins.

$$k_i = A_i \, exp\,(-E_i/RT), \quad i = 1, 2$$

For reaction 1, $P \overset{k_1}{\rightarrow} Q$,

$$k_1 = A_1 exp(-E_1/RT) \cdots \boxed{9}$$

For reaction 2, $Q \overset{k_2}{\rightarrow} R$,

$$k_2 = A_2 exp(-E_2/RT) \cdots \boxed{10}$$

Finally, we can conclude by writing down our objective function optimization (maximization) with respect to variables $t_f$ and $T$, subject to the constraints ($\boxed{1}$) to ($\boxed{10}$) above.

$$\max C_Q(t_f), \qquad s.t. \boxed{1} - \boxed{10}$$

**(b)**

Given the chemical reaction sequence below, we observe that reaction 1 (rate constant $k_1$) occurs before reaction 2 (rate constant $k_2$):

$$P \overset{k_1}{\rightarrow} Q \overset{k_2}{\rightarrow} R$$

***Considering $k_1 \gg k_2$:***
If $k_1 \gg k_2$, then reaction 1 will occur much faster than reaction 2. This results in certain key features of the plot:

- We expect to see an accumulation (hence peak) of $Q$ initially as $Q$ is produced by reaction 1 much faster than it is consumed by the next reaction (reaction 2).
- Towards later times when the reactions have progressed further, most of $P$ would have already been converted to $Q$, and the continued production of $Q$ will eventually slow to a stop when there is no more reactant $P$ left. At that time, any remaining amount of earlier-accumulated $Q$ will then be subsequently depleted by reaction 2. This gives rise to the decline in $Q$ after its peak is reached.
- When we perform the maximization optimization of our objective function, the solution or optimum point we obtain would be at a position when $Q$ is maximized, subject to all other constraints being met. The maximum amount of $Q$ such that $C_P \leq 0.0095C_{P0}$ is met, corresponds to the peak position of $Q$ as shown in the plot below. Note that at this optimum point, $C_P < 0.0095C_{P0}$ which still meets the inequality constraint on $P$.

We can sketch the expected concentration-time profiles of the three species as shown below.

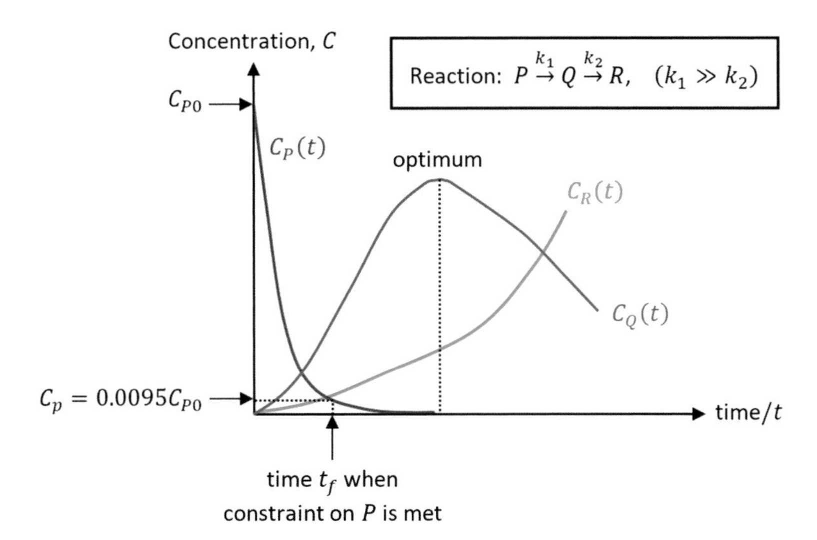


***Considering $k_1 \cong k_2$:***
Let us now consider the case whereby both reactions are occurring at almost the same rate. In this instance, we should be able to deduce some key differences in the concentration-time profile, as compared to the earlier case of $k_1 \gg k_2$:

- The accumulation of $Q$ we saw in the earlier case of $k_1 \gg k_2$ should be reduced in this case (i.e. lower peak height), since the second reaction that is consuming $Q$ is occurring just as quickly as the first reaction that produces $Q$.

- The subsequent decline in $Q$ (after its peak) should happen earlier, since reaction 2 is acting similarly quickly to consume $Q$ after it is formed from reaction 1.
- We should expect a slower decline in reactant $P$ than in the earlier case of $k_1 \gg k_2$. Since reaction 1 is less rapid with a smaller $k_1$ than before, reactant $P$ will be depleted less quickly by the reaction.
- When we perform maximization of our objective function, the solution or optimum point we obtain should again be at a position when $Q$ is maximized, subject to all other constraints being met. In this case, we observe that the optimum point has changed from the previous case.
  - The maximum amount of $Q$ such that $C_P \leq 0.0095C_{P0}$ is met, corresponds to the amount of $Q$ when $C_P = 0.0095C_{P0}$. This meets the equality part of the inequality constraint on $P$.
  - The optimum point cannot be at any earlier times than this point because that would violate the constraint $C_P \leq 0.0095C_{P0}$
  - The optimum point cannot be at any later times than this point because that would give lower amounts of $Q$, which does not fulfil our maximization objective for $Q$.

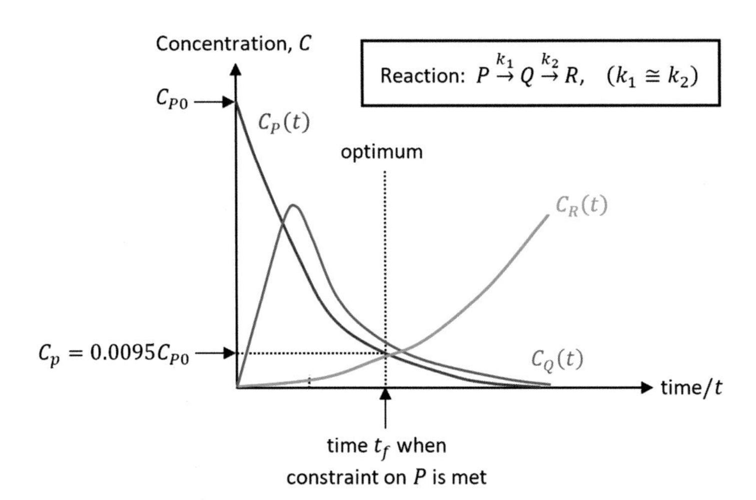


From the difference in the optimum points for each of the two cases, we can make some logical deductions in terms of implications to the profit margin for this process.

- In the first case where $k_1 \gg k_2$, there is a significant profit margin since there is a larger amount of $Q$ produced, with reactant $P$ becoming mostly converted (at least 99.05 % = (100 % − 0.95%) conversion is minimally required). More $Q$ formed leads to more final product $R$ being formed, which then leads to increased sales revenue and profit.
- In the second case where $k_1 \cong k_2$, profit margin is reduced since the two reactions are equally fast. Therefore a higher conversion of $P$ will also lead to a lower concentration of $Q$. This would ultimately lead to less amount of final product $R$ formed, and hence less profit.

**(c)**

In a batch reactor, there is no inflow and outflow of reaction mixture, therefore all of the reactant that was initially present in the reactor will simply be converted into product over time, resulting in a decrease in reactant concentration, and corresponding increase in product concentration over time.

In this problem, we are told that there is a required conversion of reactant $P$ before the reaction can be stopped at $t = t_f$, i.e. the operation duration. This requirement is such that there should be no more than 0.95% of the original amount of $P$ left, or in terms of an inequality constraint $C_P(t_f) \leq 0.0095C_{P0}$.

In order to derive an expression for the minimum value that the upper bound operation time $t_f^U$ can take, we need to identify the crossover point when the operation time is just enough to meet the required concentration of $P$ in the reactor, i.e. when $C_P$ drops to a value of $C_P(t_f)$ exactly equal to $0.0095C_{P0}$. The operation time can be longer than this minimum value but not shorter.

- If the operation time were longer than this minimum, it would be alright since a longer reaction time would deplete $P$ further, and continue to meet the required constraint of $C_P(t_f) \leq 0.0095C_{P0}$.
- However, if the operation time were any shorter than this minimum, there would be more $P$ remaining than is allowable, i.e. $C_P > 0.0095C_{P0}$ which is not possible.

Thus, we need to find an expression that would help us relate concentration of $P$ to the minimum operation time, subject to the constraint on $P$ being met. This brings us to first-order reaction kinetics.

When a reaction is first-order, it means that the rate of reaction depends linearly on only one reactant (hence order of reaction is 1 or first-order). For reactions in general, the rate of reaction $r$ can be expressed as follows (note that $r$ is a non-negative value equivalent to "the rate of formation of product" or "rate of disappearance of reactant").

$$r = \frac{dC_{product}}{dt} = \text{rate of formation of product} = -\frac{dC_{reactant}}{dt}$$
$$= \text{rate of disappearance of reactant}$$

Then for a first-order reaction where by

$$\text{Reactant} \xrightarrow{k} \text{Product}$$

The rate of reaction $r$ can be expressed as shown below, this is also known as the rate law equation and it shows the linear dependence on one reactant.

$$r = kC_{reactant}$$

Going back to our problem, we know that the relevant reaction involving $P$ is reaction 1 which is a first-order reaction, thus it should obey general first-order reaction kinetics. Hence we can write down its rate law equation:

***Reaction: 1: $P \xrightarrow{k_1} Q$***

$$r_1 = -\frac{dC_P}{dt} = k_1 C_P$$

We can derive an expression that relates concentration of $P$ with time, by integrating the above differential equations with respect to time, and defining boundary conditions for the integration where relevant. We know the initial condition at time $t = 0$, $C_P = C_{P0}$, so we can integrate from this starting point to an arbitrary time $t$ when the concentration of P is $C_P(t)$.

$$-\frac{dC_P}{dt} = k_1 C_P$$

$$\frac{dC_P}{C_P} = -k_1 dt$$

$$\int_{C_{P0}}^{C_P(t)} \frac{dC_P}{C_P} = \int_0^t -k_1 dt$$

$$[\ln C_P]_{C_{P0}}^{C_P(t)} = [-k_1 t]_0^t$$

$$\ln\left(\frac{C_P(t)}{C_{P0}}\right) = -k_1 t$$

$$C_P(t) = C_{P0} e^{-k_1 t}$$

$$\frac{C_P(t)}{C_{P0}} = e^{-k_1 t} \ldots \boxed{1}$$

This expression for $C_P$ at an arbitrary time $t$ is in fact the concentration-time profile of $P$. We observe that the amount of $P$ drops exponentially with time over the course of the reaction as it is converted into product. This is characteristic of first-order reactions.

Going back to the constraint on $P$, we know that at the end of the reactor operation when $t = t_f$,

$$C_P(t_f) \leq 0.0095 C_{P0}$$

$$\frac{C_P(t_f)}{C_{P0}} \leq 0.0095$$

We now use the earlier Eq. ($\boxed{1}$) at a specific time $t = t_f$, then substitute the expression into the inequality constraint above,

$$\frac{C_P(t_f)}{C_{P0}} = e^{-k_1 t_f} \leq 0.0095$$

$$-k_1 t_f \leq \ln 0.0095$$

$$t_f \geq \frac{-\ln 0.0095}{k_1}$$

Note that for the function of natural logarithm $\ln x$, when $x = 1$, $\ln x = 0$ and when $x < 0$, $\ln x < 0$. Therefore in the above expression, $-\ln 0.0095 > 0$, and $t_f$ is a positive value.

Now, if we want to find the minimum value for $t_f$ (or $t_f^U$), we can see from the inequality above that it should be the equality condition of the inequality,

$$\text{Minimum value of } t_f^U = -\frac{\ln 0.0095}{k_1}$$

Maximum value of $t_f^U$

The value of rate constant $k_1$ depends on temperature $T$ according to the Arrhenius' equation. Therefore the value of $k_1$ in the above expression is more precisely a function of $T$. We can express the above equation more precisely as $t_f^U = -\frac{\ln 0.0095}{k_1(T)}$ to reflect this.

If we assume that the activation energy $E_1$ for reaction 1 is positive, then by Arrhenius' equation

$$k_1 = A_1\, exp\,(-E_1/RT)$$

$$k_1 = \frac{A_1}{e^{E_1/RT}} > 0$$

We can see that if $T$ was a small value, then the denominator in the fraction $\frac{A_1}{e^{E_1/RT}}$ will be large, and the value of $k_1$ will be small. Conversely, if $T$ was a large value, then the value of $k_1$ will be large.

Since $t_f \geq -\frac{\ln 0.0095}{k_1}$, then if $k_1$ was at its smallest possible value, then $t_f$ will be at its largest possible value. The maximum upper bound $t_f^U$ can be found be setting $T$ at its minimum possible value, i.e. $T^L$ which will give the smallest value for $k_1 = k_1(T^L)$.

**(d)**

In most numerical solutions, we encounter a discretization scheme that helps "break up" a continuous function into a series of discrete parts or steps, which allows our formulated model to search systematically for the required optimum point or solution to the problem. The key features of the forward and backward difference discretization schemes are shown in the diagram below:

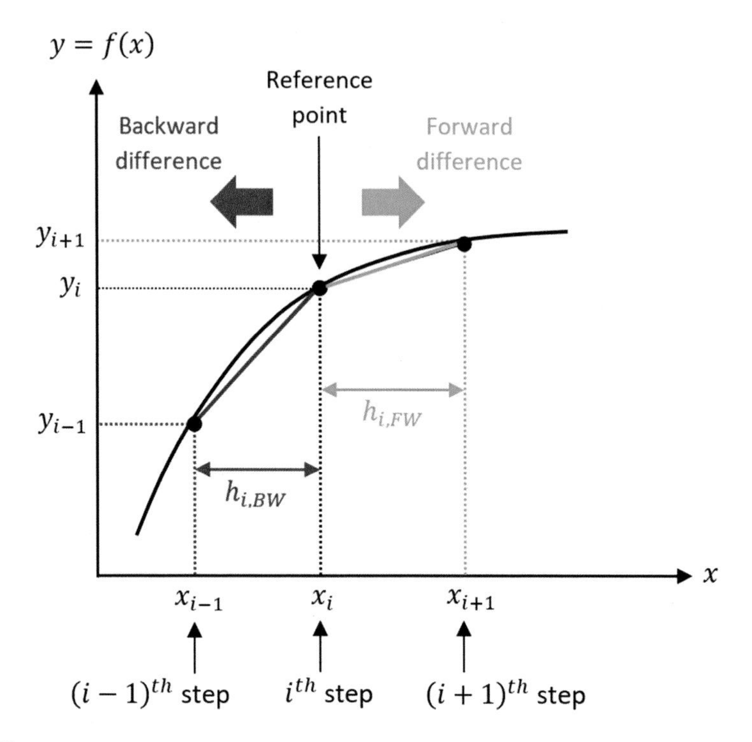


Key Features:

- The reference position is denoted $i$ or step $i$, one of the steps taken in a series of steps in the discretization scheme.
- $h_i$ is known as the "step size". In this problem, since the horizontal $x$-axis refers to time, $h_i$ is also the "time step". We note that $h_i = \Delta x$, the difference between $x$-values at the reference position and the adjacent position. Note that the step size $h_i$ may or may not be equal at each step.
  - For backward difference discretization, the adjacent position will be backward (or in the negative $x$-direction) from the reference position.
  - Conversely, for forward difference discretization, the adjacent position will be forward from the reference position.
- In each step, the non-linear continuous function $f(x)$ is approximated via a linear interpolation between the two end points of the step. The gradient of this straight line is used as an estimate of the actual gradient of the continuous function, $f'(x)$ over the span of the step. As such, we can see that the smaller the step size, the better the approximation to the continuous function, however this would also be computationally more tedious as more steps are required.

With reference to the diagram above, the forward difference method can be defined in a number of ways as follows:

$$f'(x) \cong \frac{f(x_{i+1}) - f(x_i)}{x_{i+1} - x_i} = \frac{f(x_{i+1}) - f(x_i)}{h_{i,FW}} = \frac{y_{i+1} - y_i}{x_{i+1} - x_i}$$

Similarly, the backward difference method can be defined as follows, whereby the only difference is in the direction of the step being taken:

$$f'(x) \cong \frac{f(x_i) - f(x_{i-1})}{x_i - x_{i-1}} = \frac{f(x_i) - f(x_{i-1})}{h_{i,BW}} = \frac{y_i - y_{i-1}}{x_i - x_{i-1}}$$

**(e)**

Now that we have converted our problem into a discretized one, we will need to introduce notations that introduce "steps" into the model.

Our earlier objective function took the form: max $C_Q(t_f)$. In this case, our objective function for the maximization of $Q$ remains unchanged, except that we should adjust our notations to reflect the discretization method. Hence, instead of $t_f$, we would now have max $C_{Q,\,N}$ which represents the final operation time upon taking $N$ steps in the discretization scheme.

Objective function:

$$\max \ C_{Q,N}$$

Constraints:

The initial conditions of our model remain unchanged, therefore the following constraint equations still hold for the discretized model.

$$C_P(t=0) = C_{P0} \cdots \boxed{1}$$

$$C_Q(t=0) = C_{Q0} = 0 \cdots \boxed{2}$$

$$C_R(t=0) = C_{R0} = 0 \cdots \boxed{3}$$

Next we need to include an equation that gives instructions to our model on how to proceed through the steps in this backward discretization scheme. This equation is essentially the definition of backward difference discretization in mathematical terms.

For a general function, we note that the definition of backward difference discretization can be expressed as follows.

$$f'(x) \cong \frac{f(x_i) - f(x_{i-1})}{h_{i,BW}}$$

We will now adopt this definition to our problem, whereby our function $f(x)$ now refers to concentration and $x$ refers to time.

***For species P:***
The discretization equation adapted to species $P$ for an arbitrary step $i$ can be written as follows:

$$\left.\frac{dC_P}{dt}\right|_i \cong \frac{C_{P,i} - C_{P,i-1}}{h_i}$$

From our rate law relationship established earlier in part iii, we know that

$$Reaction : P \xrightarrow{k_1} Q$$

$$\left.\frac{dC_P}{dt}\right|_i = -k_1 C_{P,i}$$

Substituting this expression into the discretization equation, we have

$$\frac{C_{P,i} - C_{P,i-1}}{h_i} = -k_1 C_{P,i}$$

$$C_{P,i} - C_{P,i-1} = -h_i k_1 C_{P,i}$$

$$C_{P,i} = C_{P,i-1} - h_i k_1 C_{P,i}, \qquad i = 1,2,\ldots N \cdots \boxed{4}$$

***For species Q:***
The discretization equation adapted to species $Q$ for an arbitrary step $i$ can similarly be written as follows:

$$\left.\frac{dC_Q}{dt}\right|_i \cong \frac{C_{Q,i} - C_{Q,i-1}}{h_i}$$

For our rate law relationship for $Q$, we need to examine all the reactions involving $Q$ which include both reactions 1 (produces $Q$) and 2 (consumes $Q$).

$$Reaction : P \xrightarrow{k_1} Q \xrightarrow{k_2} R$$

$$r = \frac{dC_{Q,i}}{dt} = r_1 - r_2 = k_1 C_{P,i} - k_2 C_{Q,i}$$

Substituting this expression into the discretization equation, we have

$$\frac{C_{Q,i} - C_{Q,i-1}}{h_i} = k_1 C_{P,i} - k_2 C_{Q,i}$$

$$C_{Q,i} - C_{Q,i-1} = h_i(k_1 C_{P,i} - k_2 C_{Q,i})$$

$$C_{Q,i} = C_{Q,i-1} + h_i(k_1 C_{P,i} - k_2 C_{Q,i}), \qquad i = 1,2,\ldots N \cdots \boxed{5}$$

***For species R:***
The discretization equation adapted to species $R$ for an arbitrary step $i$ can next be written as follows:

$$\left.\frac{dC_R}{dt}\right|_i \cong \frac{C_{R,i} - C_{R,i-1}}{h_i}$$

For our rate law relationship for $R$, since only reaction 2 involves $R$ (produces $R$), we have the following rate equation.

$$Reaction : Q \xrightarrow{k_2} R$$

$$r = \frac{dC_{R,i}}{dt} = r_2 = k_2 C_{Q,i}$$

Substituting this expression into the discretization equation, we have

$$\frac{C_{R,i} - C_{R,i-1}}{h_i} = k_2 C_{Q,i}$$

$$C_{R,i} - C_{R,i-1} = h_i k_2 C_{Q,i}$$

$$C_{R,i} = C_{R,i-1} + h_i k_2 C_{Q,i}, \qquad i = 1,2,\ldots N \cdots \boxed{6}$$

We need to further define the relationship between rate constant $k$ and temperature of reactor operation T, via the Arrhenius' equation. Hence, we have two more constraint equations as follows:

$$k_1 = A_1 exp(-E_1/RT) \cdots \boxed{7}$$

$$k_2 = A_2 exp(-E_2/RT) \cdots \boxed{8}$$

The question also indicated that the step size $h_i$ is subject to a lower bound $h^L$, therefore we will need to include this explicitly in our model as shown.

$$h_i \geq h^L, \qquad i = 1,2,\ldots N \cdots \boxed{9}$$

Next we examine if there are any required constraints on step size. The summation of all step sizes should not exceed the upper bound on final time, whereby final time or total time for reactor operation is also the result of the summation of all time steps taken. This constraint can be shown as an inequality below.

$$\sum_{i=1}^{N} h_i \leq t_f{}^U \cdots \boxed{10}$$

Finally, we need to tell the program when to stop, which refers to the requirement on the concentration of $P$ being met. Therefore our last constraint equation can be written as follows.

$$C_{P,N} \leq 0.0095C_{P0} \cdots \boxed{11}$$

Now we can conclude our model for the maximization of the objective function, i.e. max $C_{Q,\ N}$, by optimizing the values of temperature $T$, and $h_i$, and subject to constraints ($\boxed{1}$) to ($\boxed{11}$).

# Complex Optimization Problems

**Abstract** This chapter rounds up the book by consolidating the concepts and techniques covered in prior chapters, and trains the student's ability to apply these ideas to solving more challenging problems. The problems in this chapter mimic real-life scenarios, such as production planning for an industrial plant, or transportation route optimization for a logistics company, all having a desired objective (e.g. maximum profit or minimum cost) that is subject to specific constraints. In this topic, the student will gain exposure to problem types, and appreciate how the results of optimization allow key decisions to be made that would substantially drive better performance and outcomes.

**Keywords** Constrained feasible point · Constrained optimum · Non-convexity · Lagrangian function · Lagrangian multiplier · Inequality constraint · Equality constraint · Objective function · Sufficient conditions for optimality · Necessary conditions for optimality · Gradient vector · Mixing problem · Bilinear terms · Forward difference · Backward difference

## Problem 11

(a) **A non-convex function has the form below, and it is required to minimize this function using a grid-based approach.**

$$f(x_1, x_2, \ldots, x_N), \quad 0 \leq x_i \leq 1$$

**In this approach, each variable $x_i$ is discretized across $K$ number of points, and the value of the objective function $f$ is computed for all possible combinations. The minimum value out of all these combinations is then selected as a possible solution. Assuming that there is a specific precision $\varepsilon$ required in each variable, determine the total number of computations required for each of the following cases of $N$, for values of $\varepsilon = 0.1, 0.01$ and $0.001$.**



X. W. Ng, *Concise Guide to Optimization Models and Methods*,
https://doi.org/10.1007/978-3-030-84417-2_3

(i) $N = 1$
(ii) $N = 2$
(iii) $N = 5$

(b) **The objective function shown below is to be minimized subject to inequality constraints (1) to (3):**

$$\min \ (-4x_1 - 5x_2), \ \ s.t.$$

$$2x_1 - 8x_2 \leq -3 \cdots \boxed{1}$$

$$7{x_1}^3 - 15{x_1}^2 + 9x_1 + 4x_2 \leq 9 \cdots \boxed{2}$$

$$5x_1 - 5{x_2}^2 + 18x_2 \leq 30 \cdots \boxed{3}$$

**Given further that the variables $x_1$ and $x_2$ are non-negative values that shall not exceed an upper bound value of 12, determine with reasons if this problem is a convex or non-convex programming problem.**

(c) **A non-linear function $h(z)$ containing a single variable $z$ can be approximated by $\widehat{h}(z)$ using the following scheme whereby $z$ takes values between an upper and lower bound, $z^L \leq z \leq z^U$. $\gamma_i$ are free variables introduced for the approximation, and it is assumed that no more than two adjacent $\gamma_i$ can be non-zero.**

$$z = \gamma_0 \cdot z^L + \gamma_1 \cdot (k + z^L) + \gamma_2 \cdot (2k + z^L) + \ldots + \gamma_i \cdot (ik + z^L) + \ldots + \gamma_N \cdot z^U$$

$$\widehat{h}(z) = \gamma_0 \cdot h(z^L) + \gamma_1 \cdot h(k + z^L) + \gamma_2 \cdot h(2k + z^L) + \ldots + \gamma_i \cdot h(ik + z^L) + \ldots + \gamma_N \cdot h(z^U)$$

$$\gamma_0 + \gamma_1 + \gamma_2 + \ldots + \gamma_i + \ldots + \gamma_N = 1$$

$$\gamma_i \geq 0, \ \ where\ i = 0, 1, 2 \ldots N$$

**The approximation is done over a defined number of $N$ intervals, whereby the interval size $k$ can be expressed as follows.**

$$k = \frac{z^U - z^L}{N}$$

(i) **Explain how this approximation method works, and the purpose of constraints placed on the free variable, $\gamma_i$.**
(ii) **For the problem in part b, use this approximation scheme on the non-linear separable terms assuming $N = 6$. Then reformulate the problem into a linearized approximate form.**
(iii) **Discuss the relative usefulness of the methods used in part a and parts b-c.**

## Solution 11

**(a)(i)**
The number of required computations of the function increases rapidly as the precision requirement becomes more stringent (greater precision means lower values of $\varepsilon$). Let us denote the number of computations as $n$.

***Single variable N = 1:***
Before we consider the number of computations for cases of $N > 1$, let us first consider what would happen if we had only one variable in the simplest case of $N = 1$. Then we have the function,

$$f(x_1)$$

In this case, we can see that the single variable $x_1$ will have to be computed 10 times if the precision $\varepsilon = 0.1$, and this number grows exponentially to 100 for $\varepsilon = 0.01$ and 1000 for $\varepsilon = 0.001$. The number of computations $n$ is related to $N$ and $\varepsilon$ as shown below.

$$\begin{aligned} n = \left(\frac{1}{\varepsilon}\right)^N &= \left(\frac{1}{0.1}\right)^1 = 10, \ \textit{for}\ \varepsilon = 0.1 \\ &= \left(\frac{1}{0.01}\right)^1 = 100, \ \textit{for}\ \varepsilon = 0.01 \\ &= \left(\frac{1}{0.001}\right)^1 = 1000, \ \textit{for}\ \varepsilon = 0.001 \end{aligned}$$

**(a)(ii)**
***Case of N = 2:***
Now for the case of $N = 2$, the function will have two variables $x_1$ and $x_2$ instead of one.

$$f(x_1, x_2),\ \ 0 \leq x_i \leq 1$$

This means that the total number of computations will increase from the first case as follows.

$$n = \left(\frac{1}{\varepsilon}\right)^N,\ \ where\ N = 2$$

$$n = \left(\frac{1}{\varepsilon}\right)^2 = \left(\frac{1}{0.1}\right)^2 = 10^2,\ \ for\ \varepsilon = 0.1$$

$$= \left(\frac{1}{0.01}\right)^2 = 10^4,\ \ for\ \varepsilon = 0.01$$

$$= \left(\frac{1}{0.001}\right)^2 = 10^6,\ \ for\ \varepsilon = 0.001$$

**(a)(iii)**
***Case of N = 5:***
Similarly for the case of $N = 5$, the function will have five variables $x_1$, $x_2 \ldots x_5$ as follows

$$f(x_1, x_2, x_3, x_4, x_5),\ \ 0 \leq x_i \leq 1$$

This means that the number of computations will increase further as follows.

$$n = \left(\frac{1}{\varepsilon}\right)^5 = \left(\frac{1}{0.1}\right)^5 = 10^5,\ \ for\ \varepsilon = 0.1$$

$$= \left(\frac{1}{0.01}\right)^5 = 10^{10},\ \ for\ \varepsilon = 0.01$$

$$= \left(\frac{1}{0.001}\right)^5 = 10^{15},\ \ for\ \varepsilon = 0.001$$

**(b)**
This problem is a non-convex programming problem because it is subject to constraints that contain non-convex terms, such as in constraints (2) and (3).

For constraint (2), there exists a non-convex separable term marked in brackets below.

$$(7{x_1}^3 - 15{x_1}^2) + 9x_1 + 4x_2 \leq 9 \cdots \boxed{2}$$

To find out the nature (convexity/concavity) of this separable term, we can determine its second derivative as shown below.

$$\frac{d(7x_1{}^3 - 15x_1{}^2)}{dx_1} = 21x_1{}^2 - 30x_1$$

$$\frac{d^2(7x_1{}^3 - 15x_1{}^2)}{dx_1{}^2} = \frac{d(21x_1{}^2 - 30x_1)}{dx_1} = 42x_1 - 30, \ \textit{where}\ 0 \le x_1 \le 12$$

In order for the term to be fully convex, we need $\frac{d^2(7x_1{}^3-15x_1{}^2)}{dx_1{}^2} > 0$. And conversely, concavity requires $\frac{d^2(7x_1{}^3-15x_1{}^2)}{dx_1{}^2} < 0$. However, from the second derivative expression, we note that this separable term contains a concave region when $x_1 < \frac{5}{7}$, thus contributing to its non-convexity.

$$\frac{d^2(7x_1{}^3 - 15x_1{}^2)}{dx_1{}^2} = 42x_1 - 30 > 0\ (\textit{for convexity})$$

$$x_1 > \frac{5}{7}\ (\textit{convex region}); \qquad \textit{or}\ x_1 < \frac{5}{7}\ (\textit{concave region})$$

As for constraint ($\boxed{3}$), there also exists a non-convex separable term marked in brackets as shown below.

$$5x_1(-5x_2{}^2) + 18x_2 \le 30 \cdots \boxed{3}$$

We can similarly determine the second derivative of this separable term as follows.

$$\frac{d(-5x_2{}^2)}{dx_2} = -10x_2, \ \textit{where}\ 0 \le x_2 \le 12$$

$$\frac{d^2(-5x_2{}^2)}{dx_2{}^2} = \frac{d(-10x_2)}{dx_2} = -10 < 0\ (\textit{concave everywhere})$$

Since the second derivative is a constant that is always negative, this separable term is therefore concave everywhere, thus contributing to non-convexity.

**(c)(i)**

The indicated scheme performs the approximation of the non-linear univariate function, $h(z)$ through piece-wise continuous approximation over elements (or steps) in the variable $z$, in other words, this scheme acts to linearize the original function in steps, as a means to approximate it.

$$z = \gamma_0 \cdot z^L + \gamma_1 \cdot (k + z^L) + \gamma_2 \cdot (2k + z^L) + \ldots + \gamma_i \cdot (ik + z^L) + \ldots + \gamma_N \cdot z^U$$

$$\widehat{h}(z) = \gamma_0 \cdot h(z^L) + \gamma_1 \cdot h(k + z^L) + \gamma_2 \cdot h(2k + z^L) + \ldots + \gamma_i \cdot h(ik + z^L) + \ldots + \gamma_N \cdot h(z^U)$$

The number of steps taken is $N$ and each step has a size of $k$ along the $z$-axis. The upper and lower bounds established for $z$ defines the range over which the approximation is to be done.

$$k = \frac{z^U - z^L}{N}$$

The introduced free variable $\gamma_i$ acts like a "weighting factor" for each step of the approximation, and it thus makes sense that the sum of these weighting factors should equate to 1.

$$\gamma_0 + \gamma_1 + \gamma_2 + \ldots + \gamma_i + \ldots + \gamma_N = 1$$

$$\gamma_i \geq 0, \quad \textit{where } i = 0, 1, 2 \ldots N$$

$\gamma_i$ is also constrained such that no more than two adjacent $\gamma_i$ can be non-zero. This ensures that the scheme selects exactly one element of the piecewise construction and weigh its endpoints to obtain a linear interpolation within the element, as an approximation to the original non-linear function value(s).

**(c)(ii)**

To apply the scheme to part b, we need to first recognize that the variable $z$ in the approximation scheme corresponds to the variable $x_1$ (or $x_2$) in part b.

We next identify the non-linear separable terms. For constraint (2), this separable term is denoted $h_1$ while the non-linear separable term in constraint (3) is denoted $h_2$. $h_1$ and $h_2$ are functions of $x_1$ and $x_2$ respectively, presented in the format of the approximation scheme.

$$h_1(x_1) = 7{x_1}^3 - 15{x_1}^2$$

$$h_2(x_2) = -5{x_2}^2$$

From part b, we know the upper and lower bound values of $x_1$ and $x_2$,

$$0 \leq x_1 \leq 12$$

$$0 \leq x_2 \leq 12$$

Given that $N = 6$, the step size $k$ can be found as follows. The step size $k = 2$ for both functions $h_1$ and $h_2$.

$$k = \frac{x^U - x^L}{N} = \frac{12 - 0}{6}$$

$$k = 2$$

Now we can construct a series of steps with step size of 2, and compute the relevant values for the functions $h_1$ and $h_2$ as shown in the table below.

| $i$ | 0 | 1 | 2 | 3 | 4 | 5 | 6 |
|---|---|---|---|---|---|---|---|
| $x_1$ | 0 | 2 | 4 | 6 | 8 | 10 | 12 |
| $h_1(x_1) = 7x_1^3 - 15x_1^2$ | 0 | −4 | 208 | 972 | 2624 | 5500 | 9936 |
| $x_2$ | 0 | 2 | 4 | 6 | 8 | 10 | 12 |
| $h_2(x_2) = -5x_2^2$ | 0 | −20 | −80 | −180 | −320 | −500 | −720 |

The two functions, $h_1(x_1)$ and $h_2(x_2)$ contain new variables $x_1$ and $x_2$ which have been introduced in this reformulated linearized version of the original optimization problem. The reformulated problem will require two sets of variables $\gamma_i$, one for variable $x_1$ which we denote $\gamma_i|_{x_1}$ and the other for variable $x_2$ which we denote $\gamma_i|_{x_2}$.

For example in the case of $h_1(x_1)$, we have:

$$x_1 = \gamma_0|_{x_1} \cdot x_1^L + \gamma_1|_{x_1} \cdot \left(k + x_1^L\right) + \gamma_2|_{x_1} \cdot \left(2k + x_1^L\right) + \ldots + \gamma_N|_{x_1} \cdot x_1^U$$

$$\widehat{h_1}(x_1) = \gamma_0|_{x_1} \cdot h_1\left(x_1^L\right) + \gamma_1|_{x_1} \cdot h_1\left(k + x_1^L\right) + \gamma_2|_{x_1} \cdot h_1\left(2k + x_1^L\right) + \ldots + \gamma_N \cdot h_1\left(x_1^U\right)$$

The coefficients at the discrete points making up the piecewise linear approximation are shown in the table above, as $(x_1, h_1)$. And given that $N = 6$, we have

$$x_1 = \gamma_0|_{x_1} \cdot (0) + \gamma_1|_{x_1} \cdot (2) + \gamma_2|_{x_1} \cdot (4) + \ldots + \gamma_6|_{x_1} \cdot (12)$$

$$\widehat{h_1}(x_1) = \gamma_0|_{x_1} \cdot (0) + \gamma_1|_{x_1} \cdot (-4) + \gamma_2|_{x_1} \cdot (208) + \ldots + \gamma_6|_{x_1} \cdot (9936)$$

Similarly for the case of $h_2(x_2)$, we have the following formulation, again using the coefficients computed earlier in the table above as $(x_2, h_2)$, at the discrete points making up the piecewise linear approximation.

$$x_2 = \gamma_0|_{x_2} \cdot (0) + \gamma_1|_{x_2} \cdot (2) + \gamma_2|_{x_2} \cdot (4) + \ldots + \gamma_6|_{x_2} \cdot (12)$$

$$\widehat{h_2}(x_2) = \gamma_0|_{x_2} \cdot (0) + \gamma_1|_{x_2} \cdot (-20) + \gamma_2|_{x_2} \cdot (-80) + \ldots + \gamma_6|_{x_2} \cdot (-720)$$

**(c)(iii)**

The method in part a poses problems when precision requirements are high, since the number of required function evaluations is inversely proportional to precision $\varepsilon$, and hence the number of evaluations increases as precision increases with lower values of $\varepsilon$. In addition, the number of function evaluations also increases exponentially with the number of dimensions (or variables) that the problem has.

Although this method is easier to perform for simpler problems as it is applied directly to the function, the optimization process quickly becomes tedious and hence inefficient for larger problems, typical of real-life scenarios.

A better way to approximate the solution of a non-convex optimization problem is through embedded schemes such as the one in parts b-c, whereby a large number of discretizations performed across $N$ steps is able to provide a progressively high accuracy of approximation of non-convex terms.

## Problem 12

**A construction manufacturer specializes in making concrete panels and columns. These items can be sold in two forms, assembled or pre-assembled. The former is purchased by local hardware stores, while the latter is preferred by external retailers due to logistical convenience in transportation.**

**To make the concrete panels and columns, concrete mix is required as raw material and this comes in two forms, blended or unblended. The unblended form requires on-site blending at the manufacturer's own workshops before they can be further processed.**

**The raw material costs are shown in the table below.**

| Raw Material | Description | Cost ($/kg) |
|---|---|---|
| 1 | Unblended concrete mix | $c_1$ |
| 2 | Blended concrete mix | $c_2$ |

**The further processing of blended concrete includes setting and cutting, followed by assembly (for local stores only) and packing. Information on the manufacturer's processing capacities and products are shown in the tables below.**

| S/N | Product description | Customer | Selling price ($/item) | Upper bound on sales (no. of items/ month) | Raw material required (kg/item) | Processing time (mins/item) | | |
|---|---|---|---|---|---|---|---|---|
| | | | | | | Setting & Cutting | Assembly | Packing |
| 1 | Panel, assembled | Local store | $p_1$ | $x_1^U$ | $m_1$ | $t_{c1}$ | $t_{a1}$ | $t_{p1}$ |
| 2 | Column, assembled | Local store | $p_2$ | $x_2^U$ | $m_2$ | $t_{c2}$ | $t_{a2}$ | $t_{p2}$ |
| 3 | Panel, pre-assembled | External retailer | $p_3$ | None | $m_3$ | $t_{c3}$ | 0 | $t_{p3}$ |
| 4 | Column, pre-assembled | External retailer | $p_4$ | None | $m_4$ | $t_{c4}$ | 0 | $t_{p4}$ |

| Description | Value |
|---|---|
| Blending capacity (kg/month) | $B^U$ |
| Operating time for setting and cutting processes (hours/month) | $T_{cut}$ |
| Number of workers who are skilled to do either packing or assembly work | 20 |
| Number of man-hours per month per worker | 100 |

(a) **Formulate an optimization problem to plan the manufacturing process in order to maximize monthly profit.**

(b) **The delivery of products to customers can be done in a few ways. For local stores, the manufacturer will handle all deliveries directly to the stores. As for external retailers, the manufacturer can either handle the deliveries directly to the retailers' sites, or deliver to a nearby storage hub for the retailers to pick up from and continue last-mile delivery. The delivery costs for the various options are shown in the table below.**

| Product and Destination | Customer | Delivery cost ($/item) |
|---|---|---|
| Product 1 to local stores | Local stores | $d_1$ |
| Product 2 to local stores | Local stores | $d_2$ |
| Product 3 to external retailer | External retailer | $d_3$ |
| Product 4 to external retailer | External retailer | $d_4$ |
| Product 3 to storage hub (last-mile delivery handled by retailer) | External retailer | $d_5$ |
| Product 4 to storage hub (last-mile delivery handled by retailer) | External retailer | $d_6$ |

**Given that the manufacturer's own delivery operates at a maximum capacity of 1800 items per month, regardless of destination, formulate an**

**optimization problem to take into consideration the monthly delivery costs, with the same objective of maximization monthly profit. Show clearly, any new variables introduced in this problem, and how they relate to the earlier model variables in part a.**

(c) **The manufacturer is considering changing their business model in the following ways:**

(i) **Only sell to external retailers.**
(ii) **Only purchase blended concrete mix.**
(iii) **Accept an exclusivity contract from a retailer, conditional upon the manufacturer producing 250 units of each product.**

**Using the model in part a, and assuming the model parameters as shown below:**

$$t_{p3} = t_{p4} = 17$$

$$t_{c3} = t_{c4} = 32$$

$$p_3 = 22, \; p_4 = 7$$

$$m_3 = 4, \; m_4 = 1.8$$

$$c_2 = 3$$

$$T_{cut} = 520$$

**Highlight any limiting constraints that can be eased in order to help yield higher profit. Thus, recommend any improvements that can be made to the operation model.**

## Solution 12

### (a)

To begin our problem formulation, we should first define our decision variables. Let $x_i$ refer to the number of product items per month, and $r_i$ refer to the mass of raw material required in kg/month. This gives us 4 variables for $x_i$ for the 4 product types (panel, assembled; column, assembled; panel, pre-assembled; column, pre-assembled) and 2 variables for $r_i$ for the 2 types of raw material (unblended and blended concrete mix) as shown below.

$$x_i \qquad where\ i = 1, 2, 3, 4$$

$$r_j \qquad where\ j = 1, 2$$

Next we should establish all constraints to properly define our problem. First, we list down any non-negativity constraints for physical quantities.

$$x_i \geq 0 \qquad \textit{where } i = 1, 2, 3, 4$$

$$r_j \geq 0 \qquad \textit{where } j = 1, 2$$

Next, we will impose any particular constraints set by the manufacturer. First we have the upper bounds on product volumes for sales.

$$x_1 \leq {x_1}^U$$

$$x_2 \leq {x_2}^U$$

Next we have the upper bound on blending capacity for the unblended raw material.

$$r_1 \leq B^U$$

Then, we have the operating time constraint for setting and cutting. Note that there is a factor of 60 added to convert hours into minutes, for consistency in units across the inequality.

$$\sum_{i=1}^{4} t_{ci} x_i \leq 60 T_{cut}$$

Finally, we have manpower constraints for assembly and packing processes. Note that on the left-hand side of the inequality below, only products 1 and 2 require assembly, since products 3 and 4 are delivered pre-assembled. Also, there is a factor of 60 added to the right-hand side of the inequality to convert hours into minutes.

$$\sum_{i=1}^{2} t_{ai} x_i + \sum_{i=1}^{4} t_{pi} x_i \leq 20(100)(60)$$

As with most process systems, we should also consider relevant mass balances, which the model will have to consider and obey. The total mass of raw materials that is purchased should be equivalent to the total mass of raw materials contained in each finished product. Therefore, we have

$$r_1 + r_2 = m_1 x_1 + m_2 x_2 + m_3 x_3 + m_4 x_4$$

We may now conclude our model by defining our objective function $f$ that is representative of profit, and optimize this function in a maximization. Note that profit is equivalent to total sales revenue minus total cost.

$$\max_{x_i, r_j} f = \max_{x_i, r_j} \left\{ \sum_{i=1}^{4} p_i x_i - \sum_{j=1}^{2} c_j r_j \right\}$$

**(b)**

The new variables to be introduced into the model from part a should consider any additional information related to delivery options.

Let us define new variables with two new sets of subscripts. The first set of subscripts '*l*' and '*e*' indicate the customers, i.e. local stores and external retailers respectively, while the second set of subscripts '*m*' and '*r*' indicate deliveries handled by the manufacturer's own fleet only and by the retailer (for last-mile delivery) respectively.

For products 1 and 2, since they are only transported to local stores by the manufacturer's own delivery fleet, there is no need to consider the second subscript ('*m*' or '*r*') which is only relevant if the retailer handles last-mile deliveries. The first subscript ('*l*' or '*e*') is still relevant here as it reflects the customer, i.e. local stores. Hence, we have the following new variables for products 1 and 2.

$$x_{1,l} \text{ and } x_{2,l}$$

As for products 3 and 4, we should include both sets of subscripts to differentiate between the options available. For delivering to external retailers using the manufacturer's own delivery, we have the following new variables,

$$x_{3,em} \text{ and } x_{4,em}$$

As for delivering to the storage hub for last-mile delivery by the retailer, we have

$$x_{3,er} \text{ and } x_{4,er}$$

Now that we have defined our new variables, we can state any relevant constraints. Again, we shall perform a mass balance, and this gives us the following equations for the 4 products.

$$x_1 = x_{1,l}$$

$$x_2 = x_{2,l}$$

$$x_3 = x_{3,em} + x_{3,er}$$

$$x_4 = x_{4,em} + x_{4,er}$$

Now, we note that the manufacturer's own delivery operates at a maximum capacity of 1800 items per month, regardless of destination. This sets a constraint in our problem.

$$x_{1,l} + x_{2,l} + x_{3,em} + x_{4,em} \leq 1800$$

The next factor to consider is delivery cost, since this has to be included in the objective function subsequently in order to maximize profit such that it takes into account the delivery costs.

$$\text{Monthly delivery cost} = d_1 x_{1,l} + d_2 x_{2,l} + d_3 x_{3,em} + d_4 x_{4,em} + d_5 x_{3,er} + d_6 x_{4,er}$$

Note here that since $x_1 = x_{1,\ l}$ and $x_2 = x_{2,\ l}$, we can also simplify the above expression for delivery cost by replacing $x_{1,\ l}$ and $x_{2,\ l}$ with $x_1$ and $x_2$ respectively. In other words, the additional subscript '*l*' is redundant here since $x_1$ and $x_2$ unambiguously refer to products for local stores only.

$$\text{Monthly delivery cost} = d_1 x_1 + d_2 x_2 + d_3 x_{3,em} + d_4 x_{4,em} + d_5 x_{3,er} + d_6 x_{4,er}$$

Finally, we will just need to update our earlier objective function *f* in part a, by including an additional term that subtracts monthly delivery cost from total monthly revenue. This gives us a profit function that takes into account delivery costs, and a maximization of *f* will provide the required optimization solution.

**(c)**

In part c, we note that there are 3 changes to the original business model in part a. Let us consider each one, starting with the first.

If products were only sold to the external retailers, and not to the local stores, then the following equations can be established since products 1 and 2 are only sold to local stores.

$$x_{1,l} = x_1 = 0, \qquad x_{2,l} = x_2 = 0 \cdots \boxed{1}$$

Next, if only blended concrete mix is purchased as raw material, then we would have none of the unblended raw material, which gives us

$$r_1 = 0 \cdots \boxed{2}$$

Finally we note the third change which is the production requirement of 250 units of each product as a condition to exclusivity with a retailer. Therefore, we can establish that the number of each product made must be at least 250.

$$x_i \geq 250 \cdots \boxed{3}$$

Now, we will revisit the original model in part a, and re-state the portions of this original model that remain unchanged by the 3 changes.

The variables in the original model from part a were as follows:

$$x_i \qquad where\ i = 1, 2, 3, 4$$

$$r_j \qquad where\ j = 1, 2$$

Taking into account Eqs. (1) and (2), we can reduce the number of variables in our model (from what we had in part a), to just 3 variables, i.e. $x_3$, $x_4$ and $r_2$.

Next, let us look at some of our earlier constraints from part a as shown below:

$$x_i \geq 0 \qquad where\ i = 1, 2, 3, 4$$

$$r_j \geq 0 \qquad where\ j = 1, 2$$

$$x_1 \leq {x_1}^U$$

$$x_2 \leq {x_2}^U$$

$$r_1 \leq B^U$$

Since we now have only 3 variables $x_3$, $x_4$ and $r_2$, we can again simplify the above set of constraints to the following. The earlier constraints on $x_1$, $x_2$ and $r_1$ disappear since they are all equivalent to zero from (1) and (2).

$$r_2 \geq 0, \qquad x_3, x_4 \geq 0 \cdots \boxed{4}$$

Next we look at our original constraint on the operating time for setting and cutting of concrete. Since our variables have been reduced to only $x_3$ and $x_4$ for $x_i$, we can simplify the earlier constraint as shown below.

$$\sum_{i=1}^{4} t_{ci} x_i \leq 60 T_{cut}$$

Since $x_1 = x_2 = 0$,

$$t_{c3} x_3 + t_{c4} x_4 \leq 60 T_{cut} \cdots \boxed{5}$$

Similarly, our manpower constraint for assembly and packing processes can be simplified from the earlier one as follows,

$$\sum_{i=1}^{2} t_{ai} x_i + \sum_{i=1}^{4} t_{pi} x_i \leq 20(100)(60)$$

Since $x_1 = x_2 = 0$, and assembly is not required for products 3 and 4 (i.e. $t_{a3} = t_{a4} = 0$), the above expression becomes

$$t_{p3}x_3 + t_{p4}x_4 \leq 20(100)(60) \cdots \boxed{6}$$

Finally we revisit our earlier material balance constraint on the amount of concrete in the raw materials and products.

$$r_1 + r_2 = m_1x_1 + m_2x_2 + m_3x_3 + m_4x_4$$

Applying the same simplification, whereby $x_1 = x_2 = 0$ and $r_1 = 0$, we have

$$r_2 = m_3x_3 + m_4x_4 \cdots \boxed{7}$$

Finally our objective function can be revised as follows, whereby we substitute $x_1 = x_2 = 0$ and $r_1 = 0$,

$$\max_{x_i, r_j} f = \max_{x_i, r_j} \left\{ \sum_{i=1}^{4} p_i x_i - \sum_{j=1}^{2} c_j r_j \right\}$$

$$\max f = \max \{p_3x_3 + p_4x_4 - c_2r_2\} \cdots \boxed{8}$$

Let us now attempt to combine our new equations, such that we can simplify our model further. We can first substitute the expression for $r_2$ in ($\boxed{7}$) into our objective function ($\boxed{8}$) to obtain

$$\begin{aligned} \max \ \{p_3x_3 + p_4x_4 - c_2r_2\} &= \max \ \{p_3x_3 + p_4x_4 - c_2(m_3x_3 + m_4x_4)\} \\ &= \max \ \{(p_3 - m_3c_2)x_3 + (p_4 - m_4c_2)x_4\} \end{aligned}$$

We can next substitute the given parameter values for $p_3$, $p_4$, $m_3$, $m_4$ and $c_2$ into the objective function above to simplify it further. Given that $p_3 = 22$, $p_4 = 7$, $m_3 = 4$, $m_4 = 1.8$ and $c_2 = 3$, we have

$$\max \{(p_3 - m_3c_2)x_3 + (p_4 - m_4c_2)x_4\} = \max \{(22 - 4(3))x_3 + (7 - 1.8(3))x_4\}$$

$$\max f = \max \{10x_3 + 1.6x_4\} \cdots \boxed{8^*}$$

Next we can also substitute the known parameter values into constraint ($\boxed{5}$), given that $t_{c3} = t_{c4} = 32$ and $T_{cut} = 520$, we have

$$t_{c3}x_3 + t_{c4}x_4 \leq 60T_{cut} \cdots \boxed{5}$$

$$32x_3 + 32x_4 \leq 60(520)$$

$$32x_3 + 32x_4 \leq 31200$$

$$x_3 + x_4 \leq 975 \cdots \boxed{5^*}$$

Similarly, we can substitute the known parameter values into constraint ($\boxed{6}$), given that $t_{p3} = t_{p4} = 17$, we have

$$t_{p3}x_3 + t_{p4}x_4 \leq 20(100)(60) \cdots \boxed{6}$$

$$17x_3 + 17x_4 \leq 120000$$

$$x_3 + x_4 \leq 7058\ (rounded) \cdots \boxed{6^*}$$

We note here that we have two competing constraints ($\boxed{5^*}$) and ($\boxed{6^*}$) as they both refer to the sum of $x_3$ and $x_4$. Therefore, we can combine them into one by selecting the tighter constraint of the two which is the one that prevails, i.e. ($\boxed{5^*}$) as follows.

$$x_3 + x_4 \leq 975$$

From our objective function in expression ($\boxed{8^*}$), we can deduce that in order to maximize profit (or objective function $f$), we should pick the maximum limit of the inequality constraint ($\boxed{5^*}$),

$$\max f = \max\ \{10x_3 + 1.6x_4\} \rightarrow \max\ \{x_3 + x_4\} \text{ in } x_3 + x_4 \leq 975$$

This means that $x_3 + x_4 = 975$ for maximum profit.

Furthermore, since the coefficient of $x_3$ in the objective function, 10 is greater than the coefficient of $x_4$ which is 1.6, it means that the "weightage" or "contribution" of $x_3$ towards profit is greater than that of $x_4$. It follows then that we can maximize the objective function (hence profit) if we select the highest possible value for $x_3$ that would fulfill $x_3 + x_4 = 975$. Considering the extreme case if $x_3$ contributes fully to the sum of 975, then $x_4 = 0$.

$$x_3 = 975$$

$$x_4 = 0$$

However, this would violate the earlier constraint ($\boxed{3}$) we derived from the exclusivity contract which requires $x_i \geq 250$, i.e. $x_3 \geq 250$ and $x_4 \geq 250$. Therefore, we can adjust (upwards) our minimum value for $x_4$ such that it is at its minimum possible value of 250 (instead of zero which would violate ($\boxed{3}$)). With this condition in mind, we arrive at the final optimized values for maximum profit.

$$x_4 = 250$$

$$x_3 = 975 - 250 = 725$$

Substituting these optimized values back into our objective function for profit, we obtain a profit of

$$\max f = \max\ \{10x_3 + 1.6x_4\} = 10(725) + 1.6(250) = 7650$$

One of the key constraints in this problem was ($\boxed{5^*}$) as it sets the maximum limit for the number of products that could be made. And this constraint came about due to the limiting value $T_{cut}$ of 520 hours per month. Therefore, in order to obtain higher profit, this constraint could be eased by increasing the value of $T_{cut}$, i.e. increasing the operating time for setting and cutting processes. This could be done either by expanding the setting and cutting capacity per unit time, or by extending operating hours for these processes.

$$x_3 + x_4 \leq 975 \cdots \boxed{5^*}$$

Furthermore, we note that the packing time constraint ($\boxed{6^*}$) $x_3 + x_4 \leq 7058$ was redundant in the optimization, which means that there is excess capacity in packing operations which would not contribute to increased profit. By cutting down on unnecessary packing operations and channeling them to setting and cutting operations instead, there would be cost savings from the former and contributions to profit from the latter.

## Problem 13

**A pharmaceutical company produces three types of drugs 1, 2 and 3 for sale. In order to make these drugs, the company requires three different raw materials *P*, *Q* and *R*. These raw materials are blended as shown in the production scheme below, where *B* indicates the blending pool and arrows indicate material flows.**

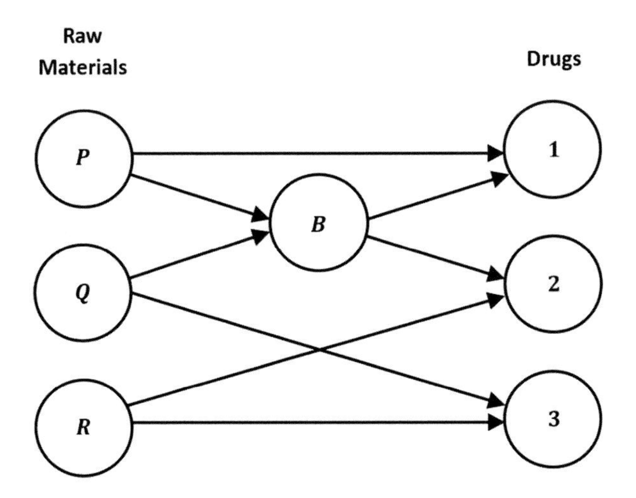

**The mass of raw materials $P$, $Q$ and $R$ used are not to exceed upper bound values of $M_P^U$, $M_Q^U$ and $M_R^U$ respectively, while the raw material content in the final products are to meet minimum levels of at least $M_1^L$, $M_2^L$ and $M_3^L$ for the three products 1, 2 and 3 respectively.**

(a) **Outline the variables involved in formulating this problem, and establish all constraints required to define the problem.**

(b) **It was found that all three raw materials contained an impurity at a concentration (mass of impurity per unit mass of raw material) denoted $C_i$ where $i = P,Q,R$. The concentration of this impurity is not to exceed an upper bound limit of $C_j^U$ in the three drug products ($j = 1,2,3$). Indicate any additional constraints that should be established in order to determine the concentration of impurity in the drug products and ensure feasibility of the problem formulation.**

(c) **The raw materials are purchased at a fixed cost (\$ per unit mass of raw material) denoted $k_i$ for $i = P,Q,R$. The drug products are sold at a price (\$ per unit mass of raw material) that can be expressed as a non-linear convex function denoted $p_j$ whereby $p_j$ is a function of the concentration of impurity $C_j$ in the product $j$ (for $j = 1,2,3$) such that when the concentration of impurity is high, the product price will be low and vice versa.**

$$p_j = f(C_j)$$

**Derive an objective function to maximize profit for the company, subject to the constraints in parts a and b. Comment, with reasons on whether this problem has a global solution.**

(d) **If the product price function $p_j$ was a constant or a linear function of $C_j$, would this problem have a definite unique global solution? Explain your answer.**

## Solution 13

**(a)**

When formulating optimal control problems, we need to first define our variables. In this problem, we have raw materials $P$, $Q$ and $R$ used to make drug products 1, 2 and 3, with an intermediate step of blending, as indicated in the production scheme. For each arrow in the scheme as shown below, we can define an associated variable $x_{ij}$ which represents the mass of raw material going from circle $i$ to circle $j$. For example, $x_{P1}$ would refer to the mass of raw material $P$ going from circle $P$ to circle 1 (product 1).

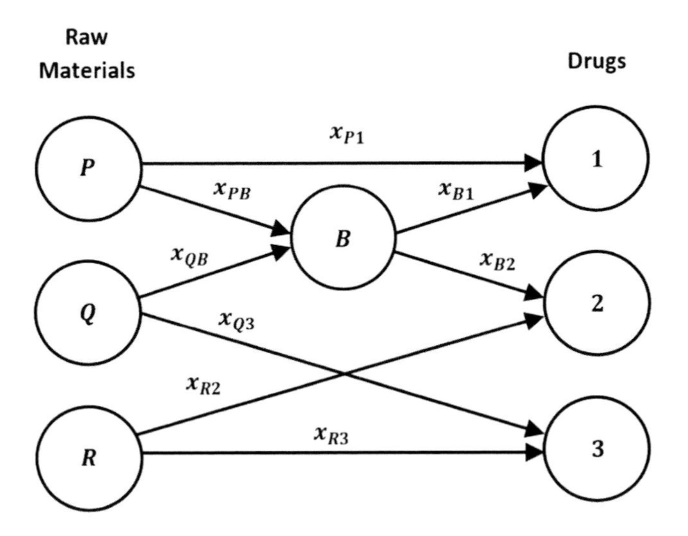


For the total number of 8 arrows, we therefore have a total of 8 variables of $x_{ij}$, whereby $i = P, Q, R, B$ and $j = 1, 2, 3, B$.

Next, we look at the constraints in our problem. As with most physical quantities, we should check for the non-negativity constraint which ensures that our solutions would be physically meaningful. In this case, we should expect non-negative values for mass, therefore we have

$$x_{ij} \geq 0 \cdots \boxed{1}$$

Next, we will look at any mass balances that should be established to ensure our system obeys laws of conservation of mass.

Let us first define our notations based on the arrows and circles in the production scheme above:

- $M_P$ refers to the total mass of raw material flowing out of circle $P$.
- $M_Q$ refers to the total mass of raw material flowing out of circle $Q$.
- $M_R$ refers to the total mass of raw material flowing out of circle $R$.

We can perform mass balances around raw materials $P$, $Q$ and $R$ respectively to obtain the following 3 constraint equations.

$$M_P = x_{PB} + x_{P1} \cdots \boxed{2}$$

$$M_Q = x_{QB} + x_{Q3} \cdots \boxed{3}$$

$$M_R = x_{R2} + x_{R3} \cdots \boxed{4}$$

Similarly, we can perform mass balances around the blending pool $B$, whereby

- $M_B$ refers to the total mass of raw material flowing into circle $B$, <u>OR</u> total mass of raw material flowing out of circle $B$.

- Total inflow is equivalent to total outflow for conservation of mass of raw materials, with no net accumulation in the blending pool.

We will arrive at the following constraint equations for material balances at *B*.

$$M_B = x_{PB} + x_{QB} \cdots \boxed{5}$$

$$M_B = x_{B1} + x_{B2} \cdots \boxed{6}$$

Finally, we can perform mass balances around the drug products 1, 2 and 3 respectively, with the following notations.

- $M_1$ refers to the total mass of raw material flowing into circle 1.
- $M_2$ refers to the total mass of raw material flowing into circle 2.
- $M_3$ refers to the total mass of raw material flowing into circle 3.

We will then obtain the next set of 3 equations as shown,

$$M_1 = x_{P1} + x_{B1} \cdots \boxed{7}$$

$$M_2 = x_{B2} + x_{R2} \cdots \boxed{8}$$

$$M_3 = x_{Q3} + x_{R3} \cdots \boxed{9}$$

Now that we have listed the constraints associated with mass balances, we will next consider any other constraints stipulated in the problem.

We note that there are upper bounds on the mass of each raw material used, therefore we can write down the following constraint, where we have added the non-negativity constraint on one side of the inequality since it applies to the raw material mass *M* at the lower limit.

$$0 \leq M_i \leq M_i^U, \qquad i = P, Q, R \cdots \boxed{10}$$

Similarly, we can write down the lower bound constraint on the raw material content required in the drug products as shown below.

$$M_j \geq M_j^L, \qquad j = 1, 2, 3 \cdots \boxed{11}$$

Finally, we arrive at our total set of 11 constraints as denoted above, to fully define this problem.

**(b)**

The additional constraints that should be established to determine the impurity concentration in the products are the ones that relate to impurity mass balance.

First, let us perform a material balance for impurity at the blending pool *B*. For this, we first need to determine the total amount of impurity entering *B*. With reference to the arrows in the production scheme, we have the following expressions for the two inflow arrows, *PB* and *QB*.

| | | | | |
|---|---|---|---|---|
| Amount of impurity along arrow $PB$ | = | Mass of raw material | $\times$ | Impurity concentration at $P$ (mass basis) |
| | = | $x_{PB}$ | $\times$ | $C_p$ |
| Amount of impurity along arrow $QB$ | = | Mass of raw material | $\times$ | Impurity concentration at $Q$ (mass basis) |
| | = | $x_{QB}$ | $\times$ | $C_Q$ |

Summing the two impurity inflow amounts, we obtain the total amount of impurity entering blending pool $B$. We also know from ($\boxed{5}$) and ($\boxed{6}$) earlier that the total inflow into $B$ is equivalent to total outflow from $B$. Therefore, we have

| | | | | | | |
|---|---|---|---|---|---|---|
| Total amount of impurity entering $B$ | | | = | Total amount of impurity leaving $B$ | | |
| Amount of impurity along arrow $PB$ | + | Amount of impurity along arrow $QB$ | = | Total mass of raw material leaving $B$ | $\times$ | Impurity concentration at $B$ (mass basis) |
| $x_{PB}C_p$ | + | $x_{QB}C_Q$ | = | $M_B$ | $\times$ | $C_B$ |

With the above, we obtain our first constraint equation as shown below.

$$M_B C_B = x_{PB} C_p + x_{QB} C_Q \cdots \boxed{12}$$

Similarly, we can perform three mass balances for each product vessel respectively. For product 1, we have

| | | | | | | |
|---|---|---|---|---|---|---|
| Total amount of impurity entering 1 | | | = | Total amount of impurity in 1 | | |
| Amount of impurity along arrow $P1$ | + | Amount of impurity along arrow $B1$ | = | Total mass of raw material entering 1 | $\times$ | Impurity concentration at 1 (mass basis) |
| $x_{P1}C_p$ | + | $x_{B1}C_B$ | = | $M_1$ | $\times$ | $C_1$ |

This gives us the next constraint equation as follows:

$$M_1 C_1 = x_{P1} C_p + x_{B1} C_B \cdots \boxed{13}$$

For product 2, we have

| | | | | | | |
|---|---|---|---|---|---|---|
| Total amount of impurity entering 2 | | | = | Total amount of impurity in 2 | | |
| Amount of impurity along arrow $B2$ | + | Amount of impurity along arrow $R2$ | = | Total mass of raw material entering 2 | $\times$ | Impurity concentration at 2 (mass basis) |
| $x_{B2}C_B$ | + | $x_{R2}C_R$ | = | $M_2$ | $\times$ | $C_2$ |

This gives us the next constraint equation as follows:

$$M_2C_2 = x_{B2}C_B + x_{R2}C_R \cdots \boxed{14}$$

For product 3, we have

| Total amount of impurity entering 3 | | | = | Total amount of impurity in 3 | | |
|---|---|---|---|---|---|---|
| Amount of impurity along arrow $Q3$ | + | Amount of impurity along arrow $R3$ | = | Total mass of raw material entering 3 | × | Impurity concentration at 3 (mass basis) |
| $x_{Q3}C_Q$ | + | $x_{R3}C_R$ | = | $M_3$ | × | $C_3$ |

This gives us the next constraint equation as follows:

$$M_3C_3 = x_{Q3}C_Q + x_{R3}C_R \cdots \boxed{15}$$

Next, we need to establish non-negativity constraints for our new variable $C_i$ since impurity concentration cannot be negative in order to be physically meaningful.

$$0 \leq C_i, \qquad i = P, Q, R, B, 1, 2, 3 \cdots \boxed{16}$$

Finally, we need the impose the upper bound limit of impurity content on the products, therefore we have

$$C_j \leq C_j^U, \qquad j = 1, 2, 3 \cdots \boxed{17}$$

The above set of additional constraints ($\boxed{12}$) to ($\boxed{17}$) would be required to fully define the problem in order to obtain feasible solutions for impurity concentrations.

**(c)**

A suitable objective function would be one that represents profit, and profit is equivalent to sales revenue minus costs. For the cost incurred in the production of the drugs, this would comprise of the total raw material cost as determined below.

$$\text{Total cost of raw materials} = k_iM_i, \quad \text{where } i = P, Q, R$$

As for the total sales revenue, this can be determined by multiplying individual product selling prices $p_j$ (in \$ per unit mass of raw material) for $j = 1, 2, 3$, with their respective sales amount (in units of mass of raw material). Therefore, we can find total revenue as follows,

$$\text{Total revenue from sale of products} = p_jM_j, \quad \text{where } j = 1, 2, 3$$

Given that the product price $p_j$ for $j = 1, 2, 3$ is a function of the concentration of impurity $C_j$ in product $j$, we can express total revenue in its functional form as shown below,

$$\text{Total revenue from sale of products} = f(C_j)M_j$$

We can now express our objective function denoted $z$ as follows, with the maximization of this function (subject to the constraints in parts a and b) giving us solutions that would yield maximum profit.

$$\max z = \max\{\sum_{j=1,2,3} f(C_j)\, M_j - \sum_{i=P,Q,R} k_i M_i\}, \qquad s.t.\ \text{constraints } \boxed{1} \text{ to } \boxed{17}$$

In order to have a unique global optimum (or solution) to this maximization problem, we will need to have a fully concave function. However, we are told that the function for product price $f(C_j)$ is non-linear and convex. Moreover, there are bilinear terms present in the objective function, such as $f(C_j)M_j$ and the terms $M_B C_B$, $M_1 C_1$, $x_{B1} C_B$, $M_2 C_2$, $x_{B2} C_B$ and $M_3 C_3$ in constraints ($\boxed{12}$) to ($\boxed{15}$) as underlined below.

$$\underline{M_B C_B} = x_{PB} C_p + x_{QB} C_Q \cdots \boxed{12}$$

$$\underline{M_1 C_1} = x_{P1} C_p + \underline{x_{B1} C_B} \cdots \boxed{13}$$

$$\underline{M_2 C_2} = \underline{x_{B2} C_B} + x_{R2} C_R \cdots \boxed{14}$$

$$\underline{M_3 C_3} = x_{Q3} C_Q + x_{R3} C_R \cdots \boxed{15}$$

These bilinear terms contribute to non-convexity in this non-linear programming problem, which means that there will be multiple feasible regions and hence multiple local optimas instead of a single unique optimum. This non-convexity arises from the blending process which creates bilinearity through unknown possible combinations of the blended raw materials, and consequently an unknown impurity content in the final products. Therefore, this problem will not yield a global solution.

**(d)**

If the product price function $p_j$ were a constant, this would not address the issue that there are bilinear terms in the constraint equations which result in non-convexity, and hence the absence of a guaranteed unique global solution.

If the price function $p_j$ were a linear function of $C_j$, bilinearity from the constraint equations still exist, and the problem (objective function and all constraint equations/inequalities to be considered) will still be non-linear and non-convex. Hence there will also not be a definite unique global solution.

## Problem 14

**In optimization, the use of Lagrange multipliers is a common method to determine the local maxima or minima of a multi-variate function subject to constraints.**

(a) **Consider the following equality constrained optimization problem whereby there are $N_{var}$ variables in the convex function $f$, and there are $N_{eq}$ equality constraints.**

$$\min_{x} f(x), \quad s.t. \quad h(x) = 0$$

**Form a Lagrangian function for this problem using the Lagrange multiplier $\lambda$, and show how it translates into a system of ($N_{var}$ + $N_{eq}$) simultaneous non-linear equations in an equal number of unknowns. Thus explain briefly how the solution to this problem can be obtained.**

(b) **Assuming that there are now $N_{ineq}$ inequality constraints added to the problem in part a as shown below, whereby $g(x)$ is a convex function.**

$$g(x) \leq 0$$

**Find an equivalent form of the Lagrangian function in this case and state the conditions required for a feasible optimum solution.**

(c) **A student claimed that the Lagrange multiplier $\lambda$ in part a "measures the sensitivity of the objective function to a perturbation in the right-hand side of the equality constraints at the optimal (or minimum) point". Do you agree with this statement? Explain using a suitable example.**

## Solution 14

**(a)**

Given the following minimization problem subject to equality constraints, we can form the Lagrangian function denoted $\mathcal{L}(x, \lambda)$ such that it is a function that varies with variables $x$ and $\lambda$ (the Lagrange multiplier for the equality constraint). Multipliers can be thought of as scalar factors applied to constraint functions.

$$\min_{x} f(x), \quad s.t. \quad h(x) = 0$$

$$\mathcal{L}(x, \lambda) = f(x) + \sum_{i=1}^{N_{eq}} \lambda_i h_i(x) = f(x) + \lambda^T h(x)$$

In the above expression, we can convert the summation term into a matrix product between the transpose of $\lambda$ and $h(x)$. For example, if we had $N_{eq} = 3$ or 3 equality constraints, we can express the three equations collectively in a $3 \times 1$ matrix or column vector, denoted $h(x)$.

$$h_1(x) = 0, \quad h_2(x) = 0, \quad h_3(x) = 0$$

$$h(x) = \begin{bmatrix} h_1(x) \\ h_2(x) \\ h_3(x) \end{bmatrix}$$

For the three constraints, we would also have 3 corresponding Lagrange multipliers. This set of multipliers can also be organized into a $3 \times 1$ matrix or column vector as shown below.

$$\lambda = \begin{bmatrix} \lambda_1 \\ \lambda_2 \\ \lambda_3 \end{bmatrix}$$

The transpose of the matrix above becomes $1 \times 3$ or a row matrix as shown below:

$$\lambda^T = [\lambda_1 \ \ \lambda_2 \ \ \lambda_3]$$

We can now clearly see how $\sum_{i=1}^{N_{eq}} \lambda_i h_i(x) = \lambda^T h(x)$ from the rules of matrix multiplication.

$$\sum_{i=1}^{N_{eq}} \lambda_i h_i(x) = \sum_{i=1}^{3} \lambda_i h_i(x) = \lambda_1 h_1(x) + \lambda_2 h_2(x) + \lambda_3 h_3(x)$$

$$\lambda^T h(x) = [\lambda_1 \quad \lambda_2 \quad \lambda_3] \begin{bmatrix} h_1(x) \\ h_2(x) \\ h_3(x) \end{bmatrix} = [\lambda_1 h_1(x) + \lambda_2 h_2(x) + \lambda_3 h_3(x)] = \sum_{i=1}^{N_{eq}} \lambda_i h_i(x)$$

The multiplication of a $1 \times 3$ matrix and a $3 \times 1$ matrix gives a resultant $1 \times 1$ matrix, i.e. a single input with value equivalent to the multiplication of each element in row 1 of the first matrix with the corresponding element in column 1 of the second matrix, to obtain the row 1 column 1 value in the final matrix. Therefore we can express our Lagrangian function in the equivalent forms as shown below,

$$\mathcal{L}(x, \lambda) = f(x) + \sum_{i=1}^{N_{eq}} \lambda_i h_i(x) = f(x) + \lambda^T h(x)$$

By taking derivatives of the Lagrangian function with respect to its variables $x$ and $\lambda$ respectively, we obtain

$$\frac{d\mathcal{L}}{dx} = \frac{df}{dx} + \sum_{i=1}^{N_{eq}} \lambda_i \frac{dh_i}{dx}$$

$$\frac{d\mathcal{L}}{d\lambda} = h^T(x)$$

We can also use the grad notations $\nabla_x$ and $\nabla_\lambda$ to represent $\frac{d}{dx}$ and $\frac{d}{d\lambda}$ respectively.

$$\nabla_x\mathcal{L} = \nabla_x f(x) + \sum_{i=1}^{N_{eq}} \lambda_i \nabla_x h_i(x)$$

$$\nabla_\lambda\mathcal{L} = h^T(x)$$

In order to determine the solution to this optimization problem, we need to locate a stationary point corresponding to the minimum value of the objective function $f(x)$ that obeys equality constraints, which means that $\nabla_x\mathcal{L} = 0$ and $\nabla_\lambda\mathcal{L} = 0$.

$$\nabla_x f(x) + \sum_{i=1}^{N_{eq}} \lambda_i \nabla_x h_i(x) = 0 \cdots \boxed{1}$$

$$h^T(x) = h(x) = 0 \cdots \boxed{2}$$

Equation ($\boxed{1}$) consists of a set of $N_{var}$ simultaneous equations, while Eq. ($\boxed{2}$) consists of a set of $N_{eq}$ simultaneous equations. Combining ($\boxed{1}$) and ($\boxed{2}$) together, we obtain a system of ($N_{var}$ + $N_{eq}$) simultaneous non-linear equations with an equal number of unknowns. This system of equations will need to be satisfied for optimality to hold, hence the solution to this system of equations would give us the optimal point or solution for minimization of the objective function, subject to its equality constraints.

**(b)**

Assuming now that we have $N_{ineq}$ inequality constraints added to the problem in part a, as shown below

$$\min_x f(x), \quad s.t. \quad h(x) = 0, \quad g(x) \le 0$$

Then our Lagrangian function will include an additional term that considers a second Lagrange multiplier for the inequality constraints, denoted $\mu$.

$$\mathcal{L}(x, \lambda, \mu) = f(x) + \sum_{i=1}^{N_{eq}} \lambda_i h_i + \sum_{i=1}^{N_{ineq}} \mu_i g_i = f(x) + \lambda^T h + \mu^T g$$

For optimality for a general non-linear programming problem, we require the gradient of the Lagrangian function (i.e. differential with respect to $x$) to be equivalent to zero at the optimum point ($x = x^*$).

$$\nabla_x \mathcal{L}(x^*, \lambda, \mu) = \nabla_x f(x^*) + \lambda^T \nabla_x h + \mu^T \nabla_x g = 0$$

In the above construction, we note the following key rules in order for a feasible optimal solution:

- The inequality multipliers must be non-negative, i.e. $\mu \geq 0$
- For the optimal point to obey the inequality constraint, $g(x^*) \leq 0$ must hold. This can be fulfilled in two ways:
  - $\mu > 0$ and $g(x^*) = 0$ – the constraint is active (i.e. the equality part of the inequality sign is in effect)
  - $\mu = 0$ and $g(x^*) < 0$ – the constraint is inactive.

The two scenarios above can be combined in a single complementarity condition as expressed below.

$$\mu_i g_i(x^*) = 0, \quad i = 1, 2..N_{ineq}$$

**(c)**

For the equality constrained problem in part a, we had the following.

$$\min_x f(x), \quad s.t. \quad h(x) = 0$$

$$\mathcal{L}(x, \lambda) = f(x) + \lambda^T h(x)$$

In order to assess the student's claim, let us consider the simplest case whereby we have only one equality constraint equation instead of a set of equalities, then $\lambda^T = \lambda$ (single value) and $h(x) = 0$ is a single equation in the expression below.

$$\mathcal{L}(x, \lambda) = f(x) + \lambda \cdot h(x) \cdots \boxed{1}$$

At the constrained minimum point, we require the following conditions to hold:

$$\left.\frac{d\mathcal{L}}{dx}\right|_{x^*} = 0, \quad \left.\frac{d\mathcal{L}}{d\lambda}\right|_{x^*} = 0$$

$$\left.\frac{d\mathcal{L}}{dx}\right|_{x^*} = \frac{df}{dx} + \lambda \frac{dh}{dx} = 0$$

$$\left.\frac{d\mathcal{L}}{d\lambda}\right|_{x^*} = h(x) = 0$$

Let us consider now that we have a slight perturbation on the right hand side of the equality constraint by an amount denoted $\varepsilon$, it then follows that $h(x)$ will no longer be zero but instead

$$h(x) = \varepsilon$$

$$h(x) - \varepsilon = 0 \cdots \boxed{2}$$

We can substitute ($\boxed{2}$) into the Lagrangian functional form ($\boxed{1}$) as shown below since the original $h(x) = 0$ is now $h(x) - \varepsilon = 0$, therefore we replace $h(x)$ with $h(x) - \varepsilon$:

$$\mathcal{L}(x, \lambda) = f(x) + \underline{\lambda \cdot (h(x) - \varepsilon)} \cdots \boxed{3}$$

The above expression is allowed to vary with the extent of perturbation $\varepsilon$, in other words $\varepsilon$ is a variable. We can differentiate the above expression with respect to $\varepsilon$ by applying the product rule on the second term as shown in underline above:

$$\frac{d\mathcal{L}}{d\varepsilon} = \frac{\partial f}{\partial x}\frac{\partial x}{\partial \varepsilon} + \underline{(h(x) - \varepsilon)\frac{\partial \lambda}{\partial \varepsilon} + \lambda \frac{\partial (h(x) - \varepsilon)}{\partial \varepsilon}}$$

$$\frac{d\mathcal{L}}{d\varepsilon} = \frac{\partial f}{\partial x}\frac{\partial x}{\partial \varepsilon} + \underline{(h(x) - \varepsilon)\frac{\partial \lambda}{\partial \varepsilon} + \lambda \left(\frac{\partial h}{\partial \varepsilon} - \frac{\partial \varepsilon}{\partial \varepsilon}\right)}$$

We can then express $\frac{\partial h}{\partial \varepsilon}$ as $\frac{\partial h}{\partial x}\frac{\partial x}{\partial \varepsilon}$ to simplify the above expression further,

$$\frac{d\mathcal{L}}{d\varepsilon} = \frac{\partial f}{\partial x}\frac{\partial x}{\partial \varepsilon} + (h(x) - \varepsilon)\frac{\partial \lambda}{\partial \varepsilon} + \lambda \left(\frac{\partial h}{\partial x}\frac{\partial x}{\partial \varepsilon} - 1\right)$$

For a small change in the value of the Lagrangian function $\Delta\mathcal{L}$,

$$d\mathcal{L} \to \Delta\mathcal{L}, \quad d\varepsilon \to \Delta\varepsilon$$

$$\Delta\mathcal{L} = \frac{\partial f}{\partial x}\frac{\partial x}{\partial \varepsilon}\Delta\varepsilon + (h(x) - \varepsilon)\frac{\partial \lambda}{\partial \varepsilon}\Delta\varepsilon + \lambda \left(\frac{\partial h}{\partial x}\frac{\partial x}{\partial \varepsilon} - 1\right)\Delta\varepsilon$$

Grouping the second and third terms together, we have

$$\Delta\mathcal{L} = \frac{\partial f}{\partial x}\frac{\partial x}{\partial \varepsilon}\Delta\varepsilon + \lambda \left(\frac{\partial h}{\partial x}\frac{\partial x}{\partial \varepsilon} - 1\right)\Delta\varepsilon$$

Taking out the common factor $\frac{\partial x}{\partial \varepsilon}$, we have

$$\Delta \mathcal{L} = \left[\frac{\partial f}{\partial x} + \lambda \frac{\partial h}{\partial x}\right] \frac{\partial x}{\partial \varepsilon} \Delta \varepsilon - \lambda \Delta \varepsilon$$

At the constrained optimum $x^*$, $\left.\frac{\partial f}{\partial x}\right|_{x^*} = 0$ and $h(x^*) = 0$ must hold, therefore $\left[\frac{\partial f}{\partial x} + \lambda \frac{\partial h}{\partial x}\right] = 0$ and we obtain

$$\Delta \mathcal{L} = [0] \frac{\partial x}{\partial \varepsilon} \Delta \varepsilon - \lambda \Delta \varepsilon$$

$$\Delta \mathcal{L}|_{x^*} = -\lambda^* \Delta \varepsilon|_{x^*}$$

$$\lambda^* = -\left.\frac{\Delta \mathcal{L}}{\Delta \varepsilon}\right|_{x^*} \cdots \boxed{4}$$

Since the equality constraint must be satisfied at the constrained optimum $x^*$, $h(x^*) = 0$. From the earlier expression ($\boxed{2}$) where we had $h(x) - \varepsilon = 0$, this also means that $\varepsilon = 0$ at the constrained optimum.

Our earlier Lagrangian function ($\boxed{3}$) was in the form:

$$\mathcal{L}(x, \lambda) = f(x) + \lambda \cdot (h(x) - \varepsilon)$$

At the constrained optimum $x^*$, $h(x^*) = 0$ and $\varepsilon = 0$, our Lagrangian function becomes

$$\mathcal{L}(x^*, \lambda^*) = f(x^*) + \lambda \cdot (h(x^*) - 0) = f(x^*) + \lambda \cdot (0 - 0)$$

$$\mathcal{L}(x^*, \lambda^*) = f(x^*)$$

Substituting this expression into our earlier expression ($\boxed{4}$)

$$\lambda^* = -\left.\frac{\Delta \mathcal{L}}{\Delta \varepsilon}\right|_{x^*}$$

$$\lambda^* = -\left.\frac{\Delta f}{\Delta \varepsilon}\right|_{x^*} \cdots \boxed{5}$$

We know that $\Delta \varepsilon$ measures the extent of perturbation from the equality condition of the equality constraint. Since $h(x) = \varepsilon$, then the following is also true

$$\Delta h = \Delta \varepsilon \cdots \boxed{6}$$

Substituting ($\boxed{6}$) into ($\boxed{5}$), we have

$$\lambda^* = -\left.\frac{\Delta f}{\Delta h}\right|_{x^*}$$

We can now take the limit whereby this perturbation is zero, to obtain the optimality condition on the Lagrange multiplier $\lambda^*$ at the constrained optimum,

$$\Delta\varepsilon \to 0$$

$$\boxed{\lambda^* = -\left.\frac{\partial f}{\partial h}\right|_{x^*}}$$

This final expression tells us that the student's claim is correct, as we can see how the Lagrange multiplier $\lambda^*$ at the constrained optimum (or minimum point) effectively measures the sensitivity of the objective function $f$ to a perturbation in the right hand side of the equality constraint.

## Problem 15

**In optimal control problems constrained by equality and/or inequality conditions, the optimal solution can be said to be a result of a 'balance of forces' between the objective function and all constraints it is subject to. Using the example of a convex objective function $f$ subject to equality and inequality constraints, show using suitable plots, what this 'balance of forces' means.**

## Solution 15

Let us assume the example of a convex non-linear objective function $f(x)$ to be minimized, subject to equality and inequality constraints as follows.

$$\min_x f(x), \quad s.t. \quad h(x) = 0, \quad g(x) \le 0$$

The optimum point occurs at a value of $x = x^*$ such that it gives a minimum value of $f = f(x^*)$, and at the same time obeys all constraints, such that $h(x^*) = 0$ and $g(x^*) \le 0$. We can think about the optimization process as a 'tension' or 'fight' between 3 forces that are going in different directions.

- First, we have the objective function that wants to be minimized.
- Next, we have equality constraints that need to be met. Only solutions with values of $x$ that obey $h(x) = 0$ are allowed.
- Finally we have the inequality constraints that also need to be met. Only solutions with values of $x$ that obey $g(x) \le 0$ are allowed.

The final solution to the optimization problem is therefore the value of $x$ that satisfies all three conditions above. In the following examples (Cases 1 and 2), we will show how this balance of forces works graphically.

**Case 1: Single-variate functions**

Consider first the simplest case whereby our objective function $f$ is a function of a single variable $x$ only, and is non-linear and convex as shown below.

$$f(x) = (x - 5)^2 - 9$$

Assume that the minimization of this objective function is subject to a linear equality constraint $h(x) = 0$, a function of a single variable $x$, as well as a linear inequality constraint $g(x) \leq 0$, also a function of a single variable $x$.

$$\min_x f(x) = \min_x \left\{(x - 5)^2 - 9\right\}$$

$$h(x) = x - 2 = 0 \cdots \boxed{1}$$

$$g(x) = 2x - 10 \leq 0 \cdots \boxed{2}$$

***Equality Constraint***

The equality constraint ($\boxed{1}$) requires that $x$ be equivalent to 2 at the constrained optimum point.

$$h(x) = x - 2 = 0$$

$$x = 2$$

We can illustrate this constraint by drawing a vertical line of $x = 2$ representing the equality constraint, such that the feasible optimum point must lie somewhere on this line in order to obey the equality constraint.

***Inequality Constraint***

As for the inequality constraint ($\boxed{2}$), we can simplify the expression by dividing both sides of the inequality by the positive number (2). Note that this does not change the directionality of the inequality sign (unlike when we divide both sides by a negative number, the directionality reverses).

$$g(x) = 2x - 10 \leq 0$$

$$2x \leq 10$$

$$x \leq 5$$

We observe that the inequality constraint requires that x only take values less than or equal to 5. This sets a feasible region to the left of the vertical line $x = 5$. It is important to note that we refer to an inequality constraint as <u>inactive</u> if the constrained optimum point lies in the region $x < 5$ and the inequality constraint is <u>active</u> if this optimum lies exactly at $x = 5$.

***Constrained Optimum***

We can represent the objective function $f(x)$ together with the constraints on the plot below against $x$. The optimum solution is said to be a result of balancing three 'forces', as it requires the consideration of (i) objective function minimization, (ii) the equality constraint, (iii) and the inequality constraint.

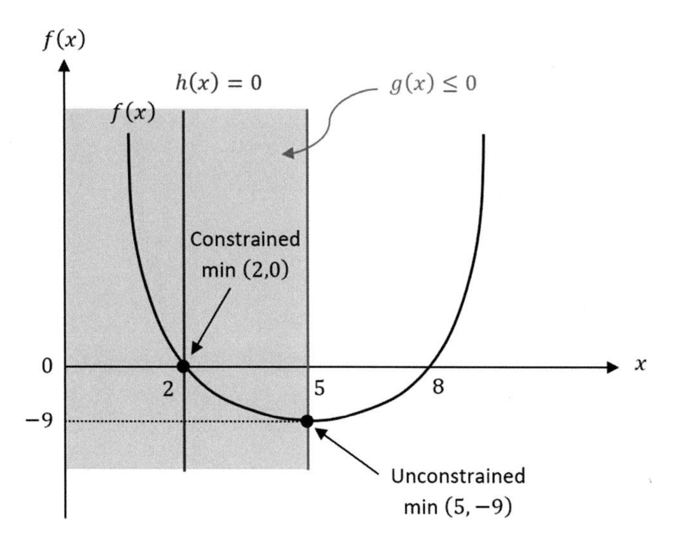


Graphically, we see that the minimum possible value of $f(x)$ that lies on a feasible point at $x = 2$ (equality constraint) and within the region $x \leq 5$ (inequality constraint) occurs at $x^* = 2$. This gives a corresponding minimum constrained value of $f(x^*) = (2-5)^2 - 9 = 0$.

If the equality constraint were absent, the minimum point (unconstrained) would have been $x^* = 5$, hence we can see the effect of the constraint on the final solution. In this instance, we also note that the inequality constraint is inactive as it does not affect the minimum point whether or not it is there, and this corresponds to when $x^* = 2 < 5$.

**Case 2: Multi-variate functions**

We have considered in Case 1 a single-variate function $f(x)$. It is common to see more complex problems involving bi-variate or multi-variate functions. In such cases, we simply add more 'dimensions' which appear as more axes in the plot. Let us now consider a minimization problem with bi-variate functions, whereby the objective function is non-linear convex, and subject to a linear equality constraint and a non-linear convex inequality constraint.

$$\min_x f(x_1, x_2) = \min_x \left\{ x_1^2 + x_2^2 - 15 \right\}$$

$$h(x_1, x_2) = x_1 - 2 + 6x_2 = 0 \cdots \boxed{1}$$

$$g(x_1, x_2) = 2x_1^2 + 0.4x_2^2 - 10 \leq 0 \cdots \boxed{2}$$

***Plot of objective function:***

$$f(x_1, x_2) = {x_1}^2 + {x_2}^2 - 15$$

In this case, the bi-variate objective function $f = f(x_1, x_2)$ includes two distinct and independent variables $x_1$ and $x_2$. It follows that the plot of $f$ will now have to be done against both the $x_1$ and $x_2$ axes, consequently making the plot 3D instead of the earlier 2D. The base area of the 3D plot is defined by the set of axes $x_1$ and $x_2$, while the height of the plot represents the value of the function $f(x_1, x_2)$. Whether a function is single-variate or multi-variate, the concept of the need to balance the forces of objective function minimization, equality constraints and inequality constraints still holds for a constrained optimisation problem.

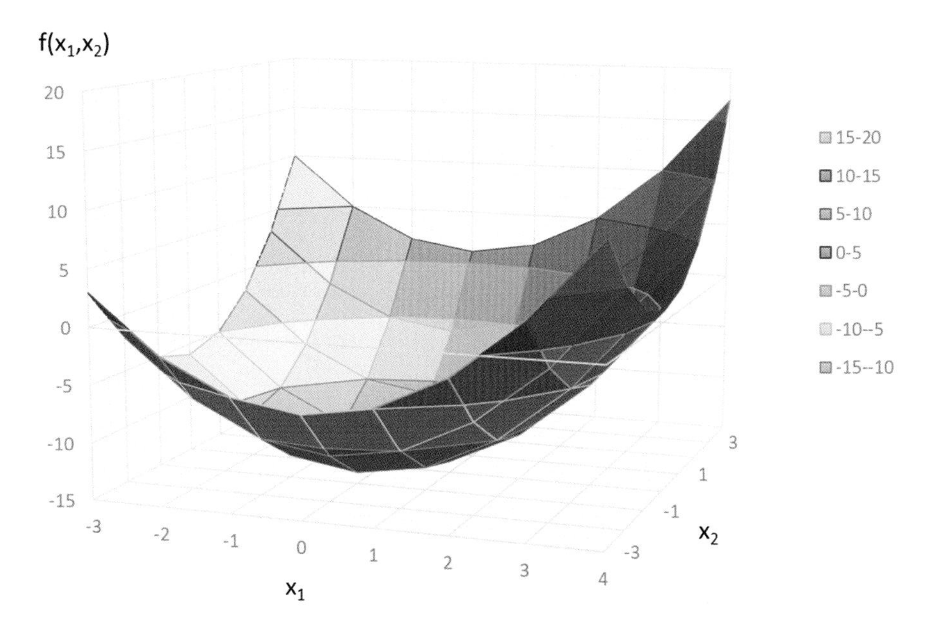


A plot of this objective function gives a 3D surface as shown below (instead of the earlier 2D line plot). Each point on this surface has an associated set of coordinates represented by $x_1$, $x_2$ and $f(x_1, x_2)$. In this plot, the different colours denote different values of $f(x_1, x_2)$ as shown in the legend on the right.

***Equality Constraint***
We can similarly have a bi-variate equality constraint as shown in (1) above.

$$h(x_1, x_2) = x_1 - 2 + 6x_2 = 0$$

When we plot the function $h(x_1, x_2) = x_1 - 2 + 6x_2$, we obtain a linear plane as shown below. Note that the orientation of this linear plane depends on the exact function, but it will be a planar surface if the function is linear in $x_1$ and $x_2$.

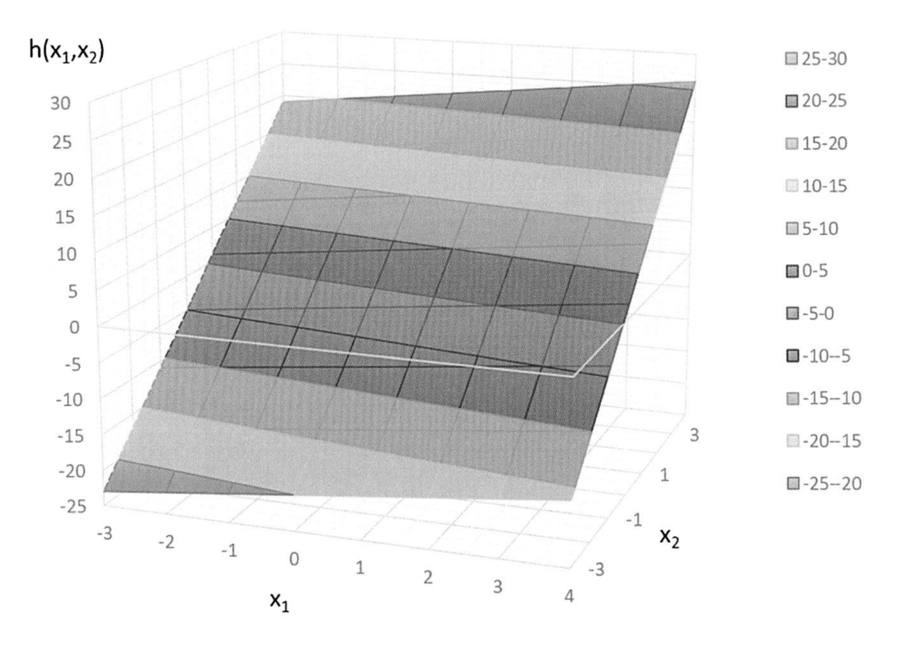


Note that the **above plot is for the function $h(x_1, x_2)$** (and not $h(x_1, x_2) = 0$) which can return any value depending on what we set for $x_1$ and $x_2$. However, in order to fulfill the equality constraint, we must define further that this function is equivalent to a particular value, in this case, zero.

$$h(x_1, x_2) = x_1 - 2 + 6x_2 = 0$$

In this case, then only points that fall on the line of intersection between the function (coloured surface representing $h(x_1, x_2) = x_1 - 2 + 6x_2$) and the horizontal plane (shaded in grey below representing $h(x_1, x_2) = 0$) will meet the equality condition. This line of intersection is described by the equation $x_1 - 2 + 6x_2 = 0$.

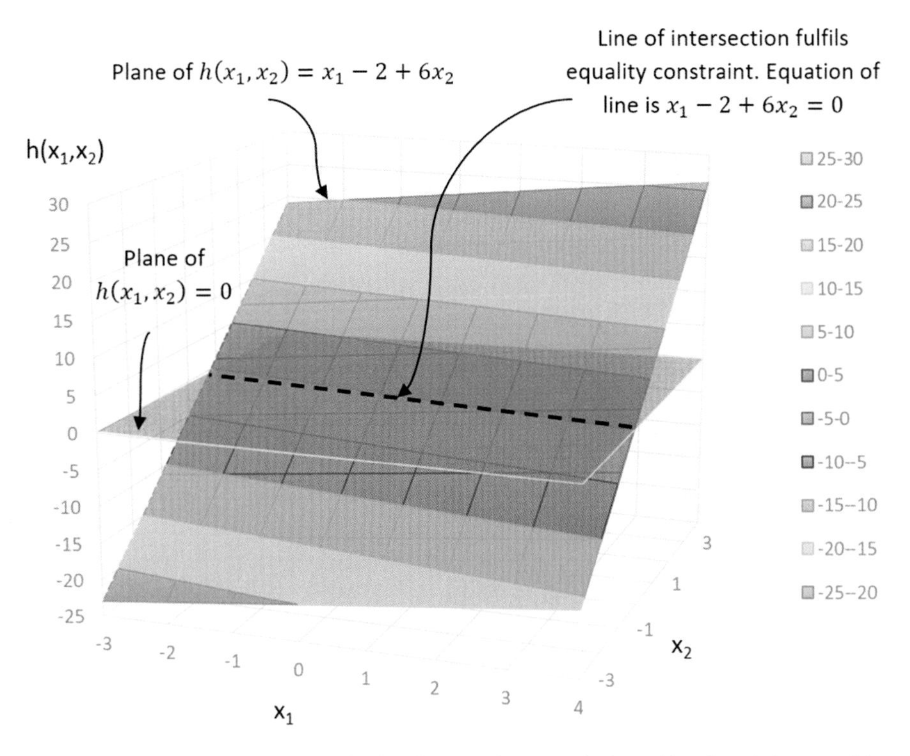


We can vertically extend this line into a plane such that all points lying on this plane will fulfill the required combination of $x_1$ and $x_2$ values that fulfill the equality condition $x_1 - 2 + 6x_2 = 0$. In other words, the equality constraint restricts in a way that for every value of $x_1$, there is a corresponding value of $x_2$ that would be automatically defined or fixed (to satisfy the equation).

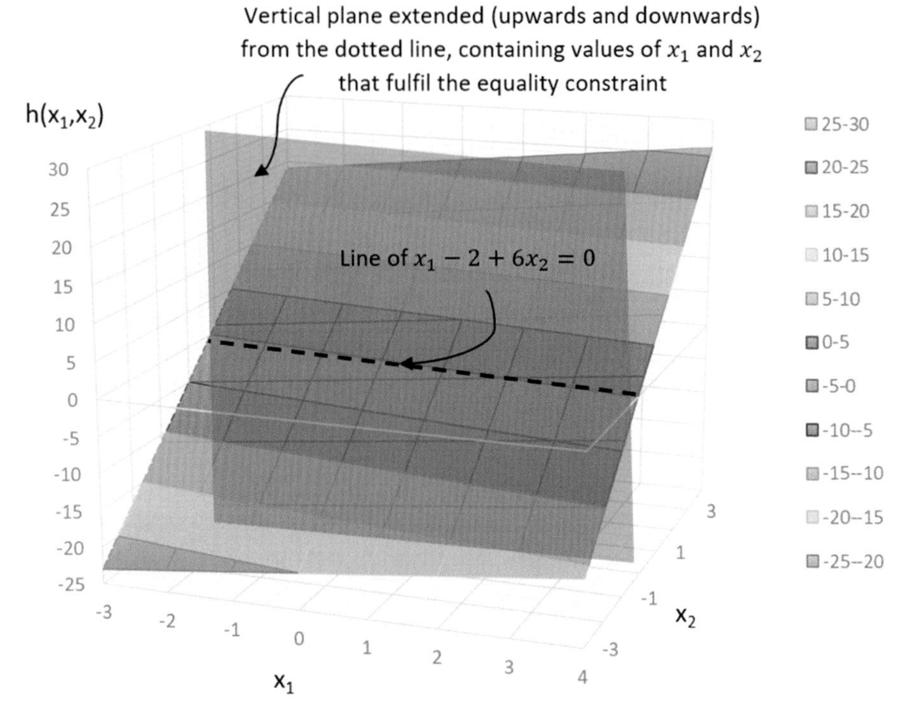


This vertical plane (shaded in blue above) will be used later when we need to find the constrained optimum point on the objective function. We also observe how the earlier vertical line that represented the equality constraint in Case 1 has now become a vertical feasible plane in Case 2 due to the additional dimension brought about by one more variable in $x$.

***Inequality Constraint***

Finally, we have our inequality constraint (2).

$$g(x_1, x_2) = 2{x_1}^2 + 0.4{x_2}^2 - 10 \leq 0$$

Similarly we observe how the earlier feasible area in Case 1 has now become a feasible volume in Case 2 due to the additional dimension brought about by one more variable in $x$. To examine this inequality constraint, we first plot the function $g(x_1, x_2) = 2{x_1}^2 + 0.4{x_2}^2 - 10$, and consider the values of this function $g$ that are $\leq 0$ to be feasible (i.e. below the horizontal plane where $g(x_1, x_2) = 0$). The feasible volume defines allowable values of $x_1$ and $x_2$ for the constrained minimum point to obey the inequality condition.

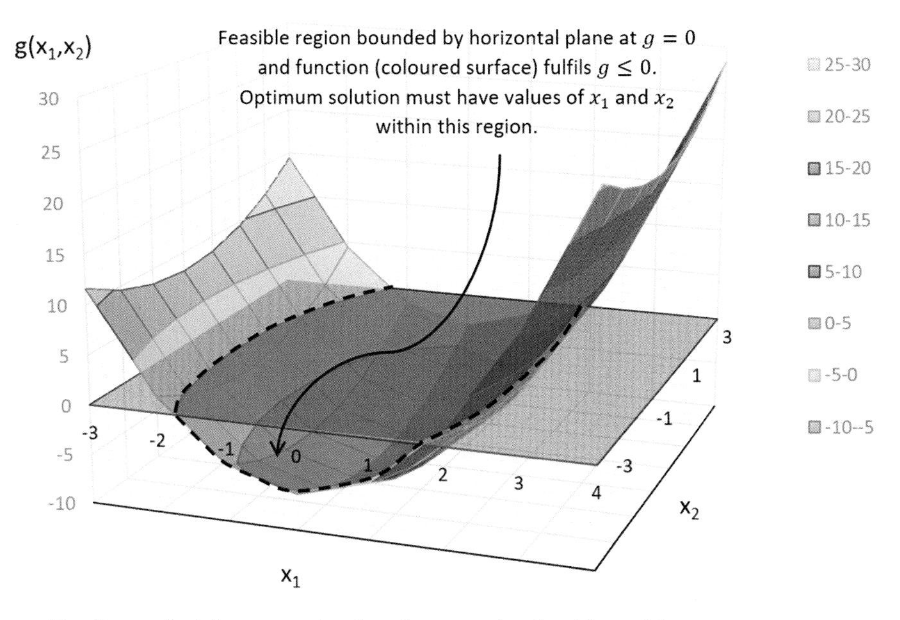


Finally, to find the constrained optimum point for this problem,

$$\min_x f(x_1, x_2) = \min_x \{x_1{}^2 + x_2{}^2 - 15\}, \qquad s.t.\boxed{1}, \boxed{2}$$

We combine the considerations of minimizing the value of $f(x_1, x_2)$, with the constraint that the point must also be located on the feasible plane (shaded blue) representing ($\boxed{1}$), and within the feasible region (shaded red) representing ($\boxed{2}$). This is shown in the plot below which presents a top-down view (i.e. contour plot) of the objective function, together with the constraints and the resultant constrained minimum point.

***Plot showing the constrained minimum:***

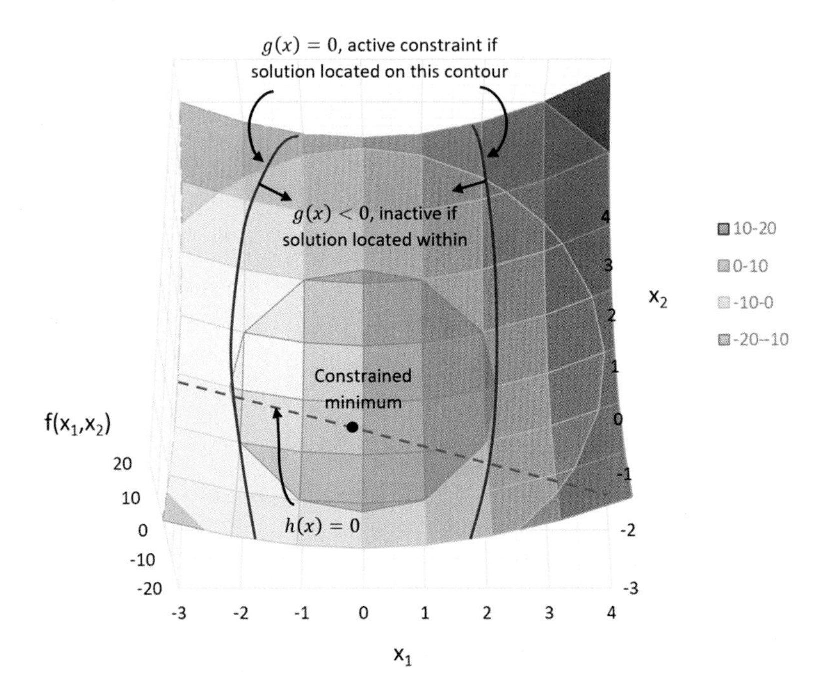


## Problem 16

**The maximization of a concave function $f(x_1, x_2)$ subject to a linear equality constraint, $h(x_1, x_2) = d$ is shown below where $a$ to $d$ are constants.**

$$\max_x f(x_1, x_2) = \max_x \{ -x_1^2 + -x_2^2 + a\} \qquad s.t. \qquad h(x_1, x_2) = bx_1 + cx_2 = d$$

(a) **Sketch the objective function above in the following two ways:**

(i) **3D profile with level-off curves shown at two arbitrary values of $f(x_1, x_2)$.**

(ii) **Top-down contour plot or 2D plan view showing the direction of the gradient of $f$ denoted $\nabla f$**

(b) **Include in your diagrams from part a, the constraint $h(x_1, x_2) = d$ such that this constraint is a plane that 'cuts through' the concave function. Then explain why the following condition holds true at the constrained maximum point, where $\lambda$ is a scalar quantity.**

$$\nabla f + \lambda \nabla h = 0$$

(c) **The Lagrangian function is commonly used to determine the constrained optimum point. Using your answer to part b, explain how the Lagrangian function works in relation to the Lagrange multiplier $\lambda$ and conditions for a feasible maximum as shown below.**

$$\mathcal{L}(x, \lambda) = f(x) + \lambda g(x), \quad s.t. \ \ g(x) = 0$$

$$\left.\frac{d\mathcal{L}}{dx}\right|_{x^*} = 0, \quad \left.\frac{d\mathcal{L}}{d\lambda}\right|_{x^*} = 0$$

(d) **Assuming now that instead of the equality constraint in parts a to c, there is an inequality constraint that is non-linear and concave. Show in your drawing how this constraint may lead to a different constrained maximum point, and what it means to have an active or inactive inequality constraint.**

## Solution 16

### (a)

We have the following concave function which is subject to a linear equality constraint.

$$\max_x f(x_1, x_2) = \max_x \left\{-x_1{}^2 + -x_2{}^2 + a\right\} \qquad s.t. \qquad h(x_1, x_2) = bx_1 + cx_2 = d$$

Let us begin by drawing the 3D profile of the objective function with the axes $x_1$ and $x_2$ forming the base, and the value of $f(x_1, x_2)$ forming the height. For a concave function, there is a maximum point (i.e. upward hill).

As shown below, the level-off curves are formed when we take a horizontal cut across the hill at a defined value of $f$. In the diagram below, we have included two level-off curves that occur at $f = k_1$ and $f = k_2$. The level-off curve is also the circumference of the exposed cut surface.

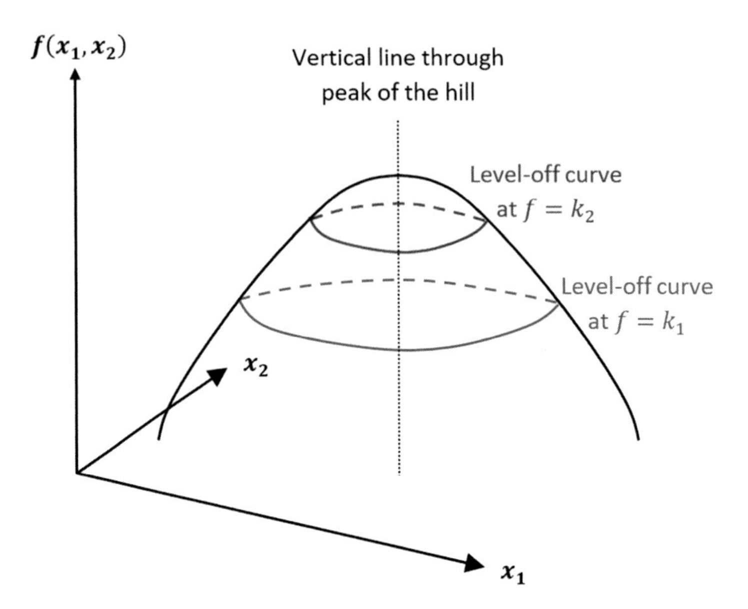


We can translate the above 3D drawing into a 2D contour plot that shows a top-down view of the hill.

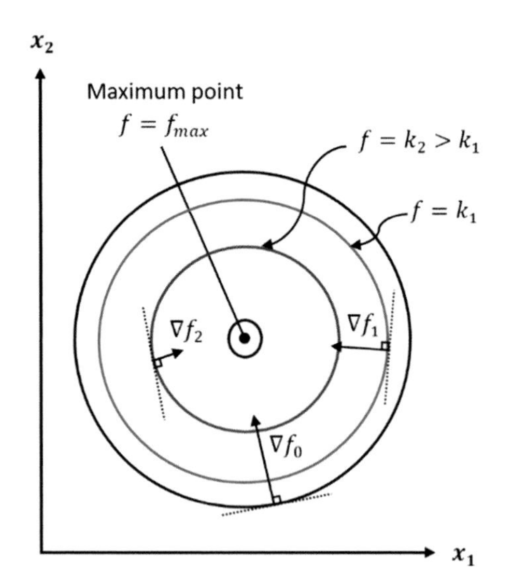


Some key features of the contour plot are indicated below:

- Inner circles represent higher values of $f$, and the value of $f$ decreases as we move radially outwards (i.e. going downhill from peak position). The $f$ axis is pointing out of the page.

- We can draw an infinite number of contours to represent the hill. In most cases, contour plots are drawn with a few representative contour circles with each one corresponding to a particular value of $f$. In the diagram above, we have drawn a few contours including the contours corresponding to the two level-off curves in our 3D drawing earlier, at heights of $f = k_1$ and $f = k_2$.
- In most cases, contour plots comprise of a successive series of circles representing values of $f$ that are at equal intervals apart. Therefore, we expect to see circles closer together near the outer region of the plot where slope is steeper than at the inner regions.
- The gradient vector for $f$ or $\nabla f$ is defined as the rate of change of $f$ with respect to $x$, and it measures how steep the slope of the hill is at a particular point. $\nabla f$ is positive as $f$ increases with $x$, i.e. positive slope or gradient (uphill).
- Since $\nabla f$ is a vector with an associated direction, it is indicated as arrows in the diagram. The arrow for $\nabla f$ (at any point on the hill) points inwards towards the peak of the hill since the value of $f$ (height) always increases as we move towards the peak for this concave function.
- The length of the arrows indicate the scalar magnitude of $\nabla f$, with the longer arrows indicating a greater 'steepness' while shorter arrows indicate gentler gradients. We observe that the slope of the hill becomes gentler as we go inwards. At the peak of the hill, $\nabla f = 0$ as the slope of the hill is horizontal at that point and therefore has zero gradient.
- The direction of $\nabla f$ is also perpendicular to the tangent to the curve at that point since it points radially inwards.

**(b)**

We can represent the equality constraint's effect on the objective function by cutting the 3D hill vertically using a plane that corresponds to the equality constraint as shown below in purple. This gives us a cut surface with a concave outline (in bold purple line), and the lower peak of this concave outline corresponds to the constrained maximum point. Note that at this constrained optimum point, the line of plane is just tangent to the curvature of the hill, i.e. the gradient of the plane is parallel to the gradient of the level-off curve at that point.

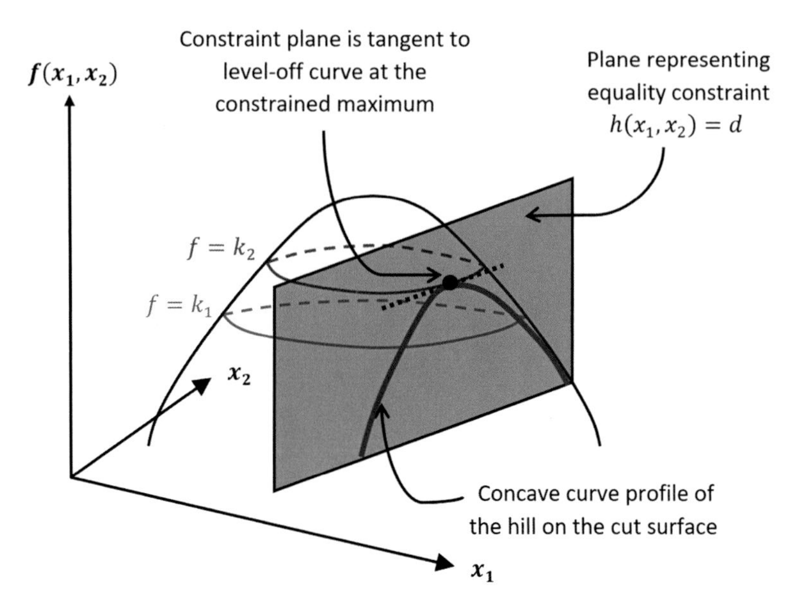


We can represent this on the contour plot as shown below. Note that in the diagram below, we have assumed an arbitrary constraint function that results in a sloped line (in purple) as shown below. The orientation and position of the plane will depend on the exact form of the function $h(x_1, x_2)$, e.g. values of $b$, $c$ and $d$.

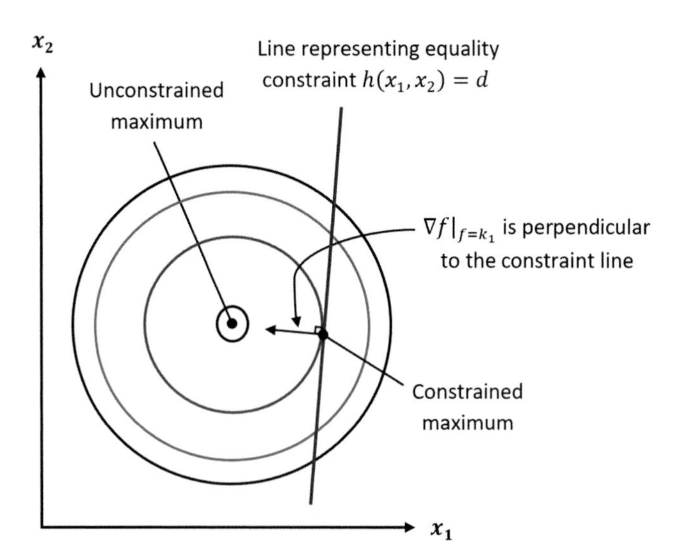


Now that we have a clearer idea of how the equality constraint comes into the picture, and how $\nabla f$ indicates the steepness of the hill's slope at a particular point, with a direction pointing radially inwards and perpendicular to the tangent at the point (in the contour plot or top-down view), let us investigate what $\nabla h$ means.

To do so, we will need to backtrack to our equality function and constraint. It is important to distinguish the constraint function from the constraint equation. The constraint function is a function $h$ that varies with $x_1$ and $x_2$. Just like any input-output mechanism of a general function, it means that the function $h$ outputs a particular value based on the input values of $x_1$ and $x_2$. Therefore, there is a range of values $h$ can take, if we provide a range of values for $x_1$ and $x_2$, and we can plot how this function looks like on a graph.

We can either plot the function $h(x_1, x_2) = bx_1 + cx_2 \cdots \boxed{1}$, or alternatively we can define our function $h$ as $h(x_1, x_2) = bx_1 + cx_2 - d \cdots \boxed{2}$. In both cases, we will be able to deduce from the graphs how the equality constraint acts. For $\boxed{1}$, the equality constraint acts as $h(x_1, x_2) = d$ while for $\boxed{2}$ the equality constraint acts as $h(x_1, x_2) = 0$. For both functions $\boxed{1}$ and $\boxed{2}$, we expect to see a linear plane tilted at some angle in 3D space which is characteristic of a linear equality function.

***Plot of function*** $\boxed{1}$
Using $h(x_1, x_2) = bx_1 - cx_2$ (assuming $b = 1$ and $c = -1$), we have

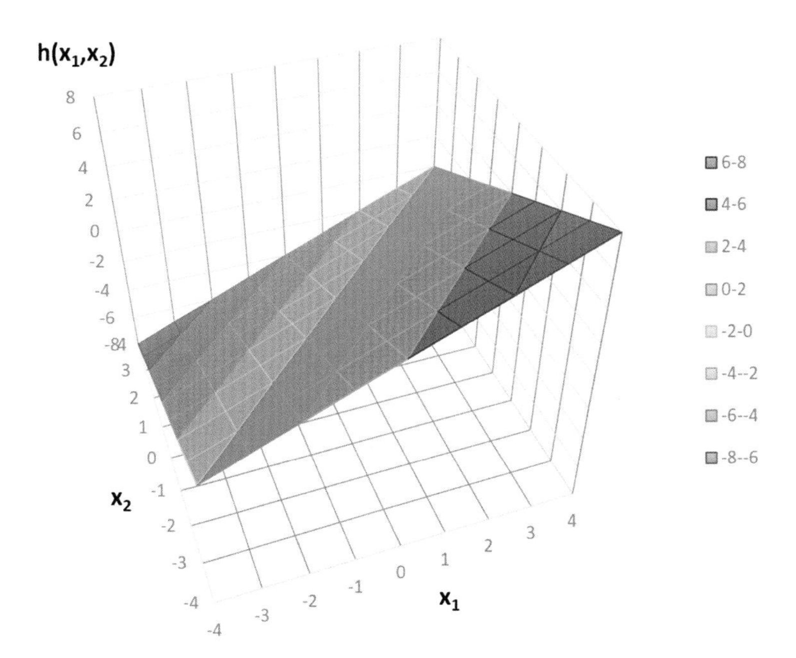


***Plot of function*** $\boxed{2}$
Using $h(x_1, x_2) = bx_1 + cx_2 - d$ (assuming $b = 1$, $c = -1$ and $d = 2$), we have the following graph instead. We notice that the function $\boxed{2}$ does not alter how the plane orientates or inclines, as it simply shifts the plane in $\boxed{1}$ downwards by 2 units.

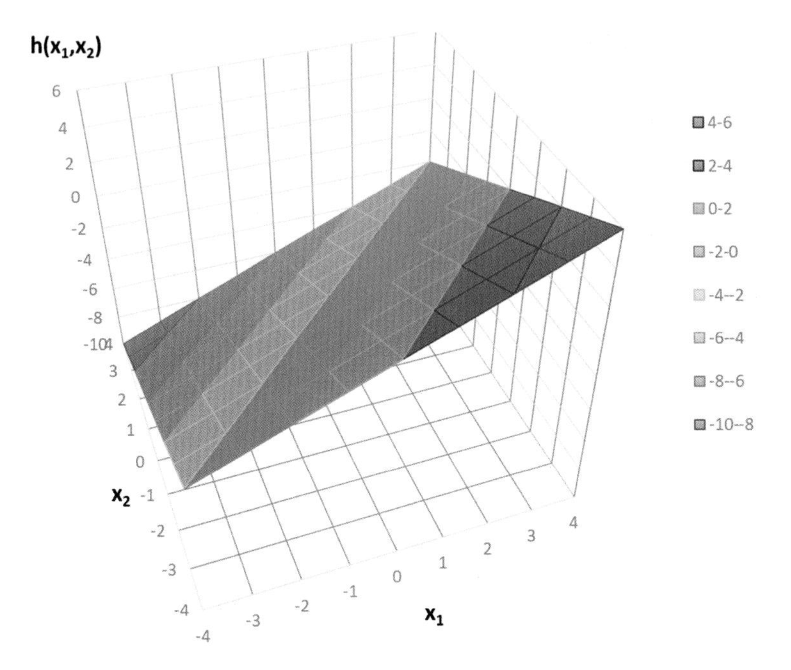


Now that we have established what our constraint functions mean, let us study what the constraint equation means. The constraint equation is simply a defined line on the function plane.

Using the example of the function $\boxed{1}$ as follows,

$$h(x_1, x_2) = bx_1 + cx_2$$

Then the constraint equation requires that the $h$ value be set to $d$, assuming $d = 2$.

$$h(x_1, x_2) = bx_1 + cx_2 = d$$
$$h(x_1, x_2) = 2$$

Graphically this translates into cutting the earlier plane horizontally at a value of $h = 2$, and the line of intersection (dotted line below) is where the equality condition is met.

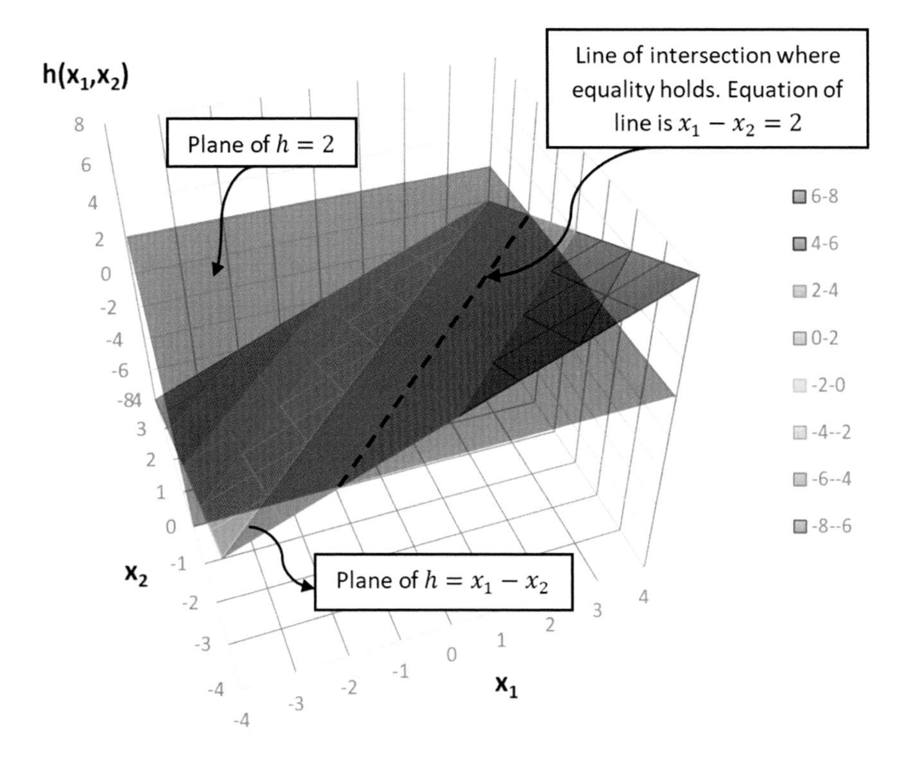


In the diagram above, we had used the example of functional form $\boxed{1}$ whereby $h(x_1, x_2) = bx_1 + cx_2$. And this plane was intersected with the horizontal plane $h(x_1, x_2) = 2$, to find the equality constraint line (dotted line of intersection). We would have obtained the same result for the equality constraint line if we had used the functional form in $\boxed{2}$ whereby $h(x_1, x_2) = bx_1 + cx_2 - d$, in which case we would have intersected it with the horizontal plane $h(x_1, x_2) = 0$ instead.

***Contour plot of the $h(x_1, x_2)$ function***

To understand what $\nabla h$ means, we need to revisit the contour plot but for the $h$ function. $\nabla h$ represents the gradient vector for the function, $h(x_1, x_2)$. In a similar way that $\nabla f$ points in a direction radially inward towards higher values of $f$ and is perpendicular to the tangent of the contour at that point, a positive value of $\nabla h$ points to higher values of $h$ and is perpendicular to each contour line (straight instead of curved lines for linear plane). If we wanted the gradient vector to point towards lower values of $h$, we would indicate it as $-\nabla h$. Using the example of the constraint function $\boxed{1}$ whereby $h(x_1, x_2) = x_1 - x_2$, we have the following contour plot for the $h$ function, showing its associated $\nabla h$ vectors, from the line of reference (line whereby the equality condition holds).

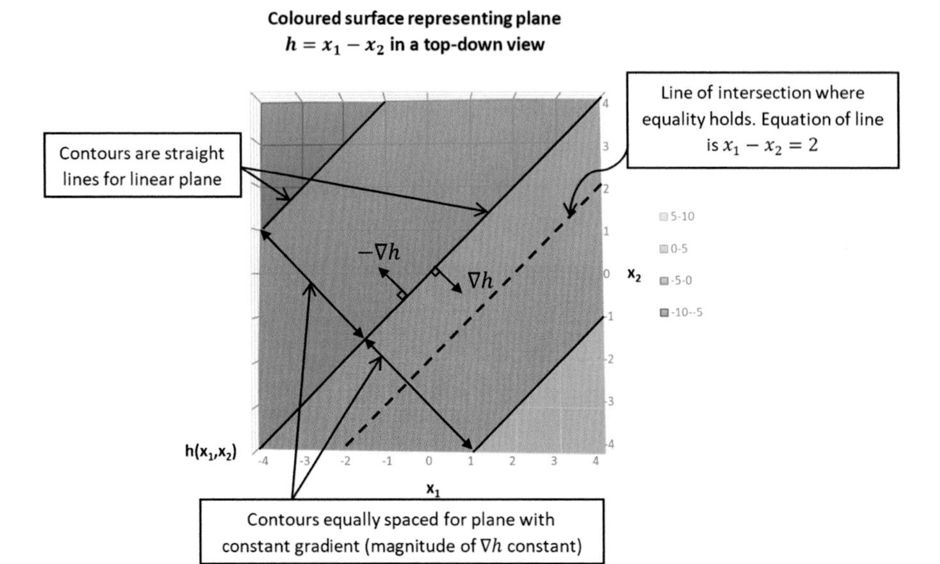


Let us now update our earlier contour plot for the objective function, by adding the contour lines (purple dotted) and gradient vector $\nabla h$ for the $h$ function as shown below.

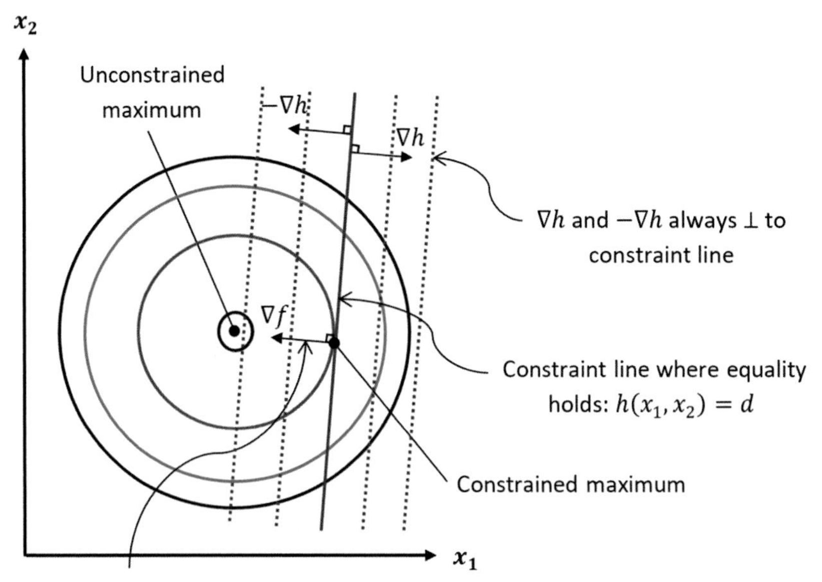

From the diagram above, we can clearly see that at the constrained optimum point, the $\nabla f$ vector is parallel to the $-\nabla h$ vector (and antiparallel to the $\nabla h$ vector). Since vectors have a magnitude and direction. While both the $\nabla f$ and $\nabla h$ vectors are parallel, their magnitudes are allowed to vary with a scalar factor $\lambda$. This makes sense since the degrees of steepness for the linear plane ($h$ function) and the concave hill ($f$ function) need not be identical at the constrained maximum point (the linear plane can 'cut' the hill at any angle, and its angle of cut depends on how sloped the plane is, which is based on the exact form of the function describing the linear plane).

This parallel relationship can be captured mathematically with the following equation if we assume $\lambda > 0$.

$$\text{At the constrained maximum}: \quad \nabla f = \pm\lambda\nabla h$$

The inclusion of a $\pm$ sign, is to indicate that the parallel relationship encompasses the anti-parallel case when the $\nabla h$ and $\nabla f$ vectors are pointing in directly opposite directions but are still parallel, hence

$$\nabla f = -\lambda\nabla h$$

**(c)**

From the analysis in part b, we note that at the constrained optimum point, the two gradient vectors $\nabla f$ and $-\nabla h$ are parallel. Assuming that $\lambda$ takes positive values, then as stated in part b before, the following must be true.

$$\nabla f(x) = -\lambda\nabla h(x) \cdots \boxed{1}$$

At the constrained optimum, the equality condition must also be met,

$$h(x) = d \cdots \boxed{2}$$

In order to fit the form of the Lagrangian function given in the problem $\mathcal{L}(x,\lambda) = f(x) + \lambda g(x)$, which is subject to an equality of the form $g(x) = 0$, we can re-express our Eq. ($\boxed{2}$) into the required form as follows.

$$g(x) = h(x) - d = 0$$

Now we can substitute our expression for $g(x)$ into the Lagrangian function. Also noting that we have two variables in $x$, $x_1$ and $x_2$, we can express our Lagrangian function as follows:

$$\mathcal{L}(x_1,x_2,\lambda) = f(x_1,x_2) + \lambda g(x_1,x_2)$$

$$\mathcal{L}(x_1,x_2,\lambda) = f(x_1,x_2) + \lambda(h(x_1,x_2) - d)$$

Or more simply,

$$\mathcal{L} = f + \lambda(h - d)$$

The Lagrangian function is particularly useful in solving optimisation problems, as it can be used to find the constrained optimum point at $x = x^*$ by subjecting the function to conditions ($\boxed{a}$) and ($\boxed{b}$) as shown below.

$$\left.\frac{d\mathcal{L}}{dx}\right|_{x^*} = 0 \cdots \boxed{a}$$

$$\left.\frac{d\mathcal{L}}{d\lambda}\right|_{x^*} = 0 \cdots \boxed{b}$$

Now notice how the two conditions ($\boxed{a}$) and ($\boxed{b}$) exactly satisfy the earlier two requirements ($\boxed{1}$) and ($\boxed{2}$) on having parallel (or antiparallel) gradient vectors $\nabla f$ and $\nabla h$, and satisfy the equality $h(x) = d$.

When we set $\left.\frac{d\mathcal{L}}{dx}\right|_{x^*} = 0$, we have

$$\mathcal{L} = f + \lambda(h - d)$$

$$\frac{d\mathcal{L}}{dx} = \nabla_x \mathcal{L} = \nabla_x f + \lambda \nabla_x h = 0$$

$$\nabla_x f = -\lambda \nabla_x h$$

The above is simply requirement ($\boxed{2}$).

And when we set $\left.\frac{d\mathcal{L}}{d\lambda}\right|_{x^*} = 0$, we have

$$\frac{d\mathcal{L}}{d\lambda} = \nabla_\lambda \mathcal{L} = h - d = 0$$

$$h = d$$

The above is simply requirement (2).

As we can see, the Lagrangian function is cleverly designed to capture the requirements of optimisation. By setting the first derivative of the Lagrangian function with respect to its variables ($x_1$, $x_2$, $\lambda$) to zero, we are effectively trying to locate the maximum (or minimum) constrained point that has to correspond to a point whereby the gradient vectors of the objective function and the constraint function are equal.

**(d)**

If we had an inequality constraint that is non-linear and concave, denoted as $g(x_1, x_2) \geq 0$, then we can consider two cases,

- Equality condition of the inequality constraint $g(x_1, x_2) = 0$. This is known as an active inequality constraint.
- Inequality condition of the inequality constraint $g(x_1, x_2) > 0$. This is known as an inactive inequality constraint.

We can show in the contour plot below how this inequality constraint acts. In the absence of any constraints, the objective function will simply have its maximum point at the maximum value of the concave function (i.e. peak of the hill in the side view diagram). When we apply a constraint however, this maximum value is "limited" in that it takes a lower value than the original unconstrained maximum value. Therefore, we can visualise the effect of constraints as tending to shift the optimum point in a direction away from its optimum position when unconstrained.

**Diagram 1: Active constraint**

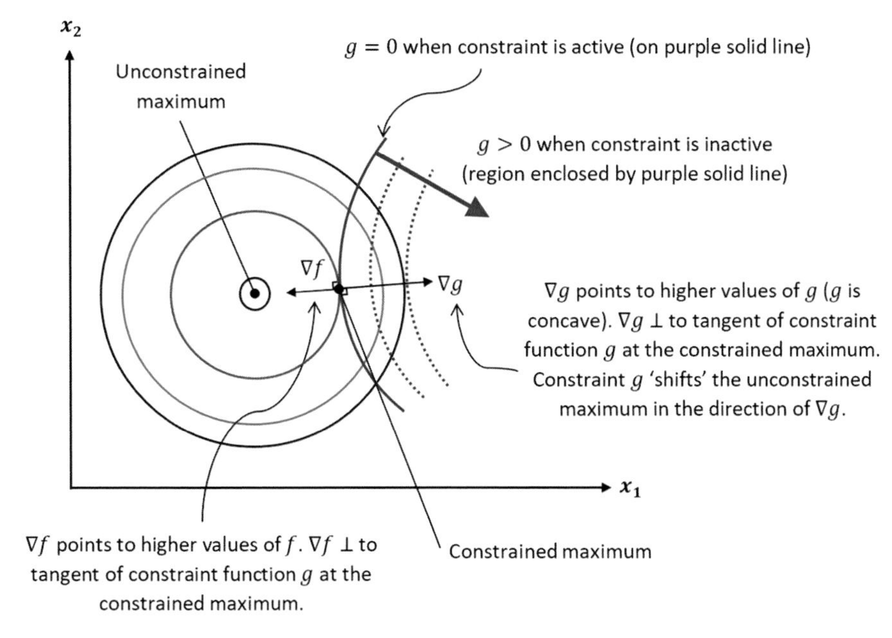


In the diagram above, the constraint is active, since the constrained maximum lies exactly on the purple solid line where the equality part of the inequality constraint holds, i.e. $g = 0$. At this point, the two curves (red and purple) just touch each other at a single point, which is the constrained optimum point.

If the constraint was inactive, then the constrained maximum point would lie within the region $g > 0$ instead, as shown below. This means that the constraint's effect is "not felt", and hence does not affect the position of the optimum point whether or not it is there. It therefore makes sense that the unconstrained maximum will then be the same as the constrained maximum for an inactive inequality constraint in this problem (in the absence of other constraints).

**Diagram 2: Inactive constraint**

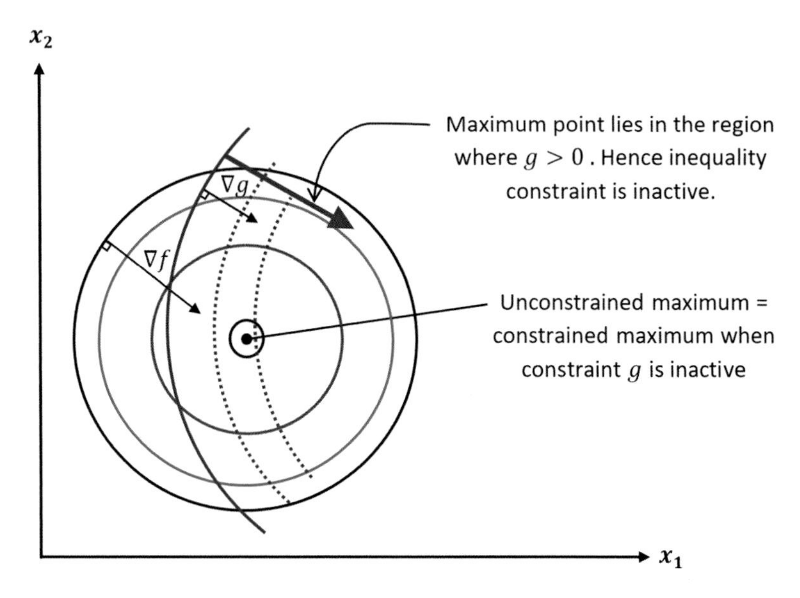


Now, let us assign a Lagrange multiplier $\mu$ to the inequality constraint whereby $\mu \geq 0$ and rewrite the Lagrangian function as follows.

$$\mathcal{L} = f + \mu g$$

Consider the case when the **constraint is inactive ($\mu = 0$)**, then it follows that the constraint acts like it is not there, and does not change the optimum point position from the unconstrained state. We know that for any optimum point (maximum or minimum), it must be a stationary point such that $\nabla_x f = 0$.

$$\mathcal{L} = f + \mu g, \ \ \mu = 0$$

$$\mathcal{L} = f$$

At stationary point, $\nabla_x f = 0$

$$\frac{d\mathcal{L}}{dx} = \nabla_x \mathcal{L} = \nabla_x f = 0$$

Now let us consider what happens when the **constraint is active ($\mu > 0$)**, and $g = 0$. This means that we simply return to the earlier case in part b, where we had an objective function subject to an equality constraint. And it therefore follows that when we set $\left.\frac{d\mathcal{L}}{dx}\right|_{x^*} = 0$, we have the following,

$$\mathcal{L} = f + \mu g, \quad \mu > 0$$
$$\nabla_x \mathcal{L} = \nabla_x f + \mu \nabla_x g = 0$$
$$\nabla_x f = -\mu \nabla_x g$$

The above corresponds to the requirement for a constrained maximum, whereby the gradient vectors $\nabla_x f$ and $\nabla_x g$ are parallel, and pointing in directly opposite directions. This is also shown in diagram 1 above.

When we set the second condition $\left.\frac{d\mathcal{L}}{d\mu}\right|_{x^*} = 0$ at the constrained optimum, we have

$$\frac{d\mathcal{L}}{d\mu} = \nabla_\mu \mathcal{L} = g = 0$$
$$g = 0$$

The above expression then also corresponds to the second requirement that the constrained maximum point lies on the purple bold line where $g = 0$, equality condition is fulfilled, or we may say that the inequality constraint is active.

**Complementarity condition:**
We may summarise our results from the above analysis for an active and inactive inequality constraint, by casting it in a complementarity condition or property about inequality constraints.

The complementarity condition states that for an inequality constraint with a non-negative Lagrange multiplier $\mu \geq 0$, the following must be true:

$$\mu g(x^*) = 0$$

The above expression means that at the constrained optimum point $x^*$, either $\mu = 0$ or $g(x^*) = 0$. In the former case the inequality constraint is said to be inactive, and we have the constrained optimum point lying in the region where $g > 0$. In the latter case, $\mu > 0$ and $g(x^*) = 0$. The inequality constraint is said to be active and it is equivalent to an equality constraint since the equality part of the inequality sign holds true. The complementarity condition captures in a single equation both possibilities for the inequality constraint $g \geq 0$ at the constrained optimum point.

**Summary of necessary conditions for optimality**
We can combine our findings in part b (for equality constraints) and part c (for inequality constraints) and summarise them into an overall set of conditions required for a constrained optimum point.

Maximization of a concave function
For the maximization of a concave objective function $f$, subject to an equality constraint $g_1 = 0$ and a concave inequality constraint $g_2 \geq 0$,

$$\max_x f \qquad s.t. \qquad g_1 = 0, \;\; g_2 \geq 0$$

The Lagrangian function $\mathcal{L}$ can be written in the form below, where $\lambda$ and $\mu$ are non-negative values representing the Lagrange multipliers for the equality and inequality constraint functions respectively.

$$\mathcal{L} = f + \lambda g_1 + \mu g_2$$

The necessary conditions to be met at the constrained maximum, $x = x^*$ are:

1. $\nabla_x \mathcal{L} = 0 \;\rightarrow$ Gradient vector of objective function is parallel to the gradient vector of all active constraints. Active constraint(s) being either cases a or b below (note how the equality constraint is always active):
   (a) Active equality constraint $g_1$ only (if $g_2$ were inactive which means $g_2|_{x^*} > 0$)
   (b) Active equality constraint $g_1$ and active inequality constraint $g_2$ (whereby equality part of the inequality sign holds $g_2|_{x^*} = 0$).
2. $\nabla_\lambda \mathcal{L} = 0 \rightarrow$ Equality constraint is satisfied, $g_1|_{x^*} = 0$.
3. $\nabla_\mu \mathcal{L} = 0 \;\rightarrow$ Inequality constraint (active) is satisfied as an equality $g_2|_{x^*} = 0$. (If the inequality constraint were inactive, $\mu = 0$ and the third term "$\mu g_2$" drops out from the Lagrangian function.)

Minimization of a convex function

We can reverse the above requirements on a concave function maximization, to formulate the corresponding required set of conditions for the minimization of a convex function *f*, subject to an equality constraint $g_1 = 0$ and a convex inequality constraint $g_2 \leq 0$,

$$\min_x f \qquad s.t. \qquad g_1 = 0, \;\; g_2 \leq 0$$

The Lagrangian function can again be written in the form as follows

$$\mathcal{L} = f + \lambda g_1 + \mu g_2$$

And the necessary conditions to be met at the constrained minimum, $x = x^*$ are:

1. $\nabla_x \mathcal{L} = 0 \;\rightarrow$ Gradient vector of objective function is parallel to the gradient vector of all active constraints.
2. $\nabla_\lambda \mathcal{L} = 0 \rightarrow$ Equality constraint is satisfied, $g_1|_{x^*} = 0$.
3. $\nabla_\mu \mathcal{L} = 0 \;\rightarrow$ Inequality constraint (active) is satisfied as an equality $g_2|_{x^*} = 0$. (If the inequality constraint were inactive, $\mu = 0$ and the third term "$\mu g_2$" drops out from the Lagrangian function.)

# Index





X. W. Ng, *Concise Guide to Optimization Models and Methods*,
https://doi.org/10.1007/978-3-030-84417-2

Printed in the United States
by Baker & Taylor Publisher Services